Light and Electron Microscopy of Cells and Tissues

An Atlas for Students in Biology and Medicine

Light and Electron Microscopy of Cells and Tissues

An Atlas for Students in Biology and Medicine

Edmund B. Sandborn, M.D.

Département d' Anatomie   Faculté de Médecine   Université de Montréal   Montréal, Canada

ACADEMIC PRESS   New York and London

ACADEMIC PRESS, INC.
111 Fifth Avenue, New York, New York 10003

United Kingdom Edition published by
ACADEMIC PRESS, INC. (LONDON) LTD.
24/28 Oval Road, London NW1 7DD

LIBRARY OF CONGRESS CATALOG CARD NUMBER: 79-188864

PRINTED IN THE UNITED STATES OF AMERICA

TABLE OF CONTENTS

PREFACE

Recent advances in fixation of tissues for electron microscopy have resulted in the preservation of a greater detail of fine structure than was previously anticipated. In addition they permit a better preservation of tissues for routine examination of control sections in the light microscope. Adequately differentiated, toluidine blue-stained sections of one micron or less in thickness, when examined in the light microscope, allow a considerable improvement in resolution over conventionally prepared sections. As a consequence, it was decided that a coordinated study by both light and electron microscopy of thin sections of the most commonly encountered tissues of animals would be of value to the student of biological structure. Some aspects of cell and organelle surfaces, obtained by the freeze-etching technique, have been included to provide a tridimensional concept of the cell. Functional aspects have been considered wherever applicable and, in this respect, a few illustrations of the results of histochemical, cytochemical, and autoradiographic studies have been included. Unless otherwise stated, the tissues have been taken from the perfused laboratory rat.

This atlas is a condensation of Volumes I and II of "Cells and Tissues by Light and Electron Microscopy" (Academic Press, Inc., New York and London, 1970). It is designed to provide the student with the essentials for a basic knowledge of animal cells and tissues.

EDMUND B. SANDBORN

ACKNOWLEDGMENTS

The experimental work which made this book possible was supported by grants from the Medical Research Council of Canada, the Muscular Dystrophy Association of Canada, the National Cancer Institute of Canada, the Quebec Heart Foundation, and the Quebec Research Council. For the laboratory space and services and for routine maintenance I am deeply indebted to the Université de Montréal.

I acknowledge with appreciation permission to reproduce figures from the following publications: the *Journal of Ultrastructure Research*, the *Canadian Journal of Physiology and Pharmacology*, and the *Revue Canadienne de Biologie*. Figures 1 (6) and 2 (64) are reproduced by permission from the *Journal of Ultrastructure Research* **11**, 123–138 (1964) and **18**, 695–702 (1967); Figure 1 (64) is reproduced by permission of the National Research Council of Canada from the *Canadian Journal of Physiology and Pharmacology* **44**, 329–338 (1966); Figures 1 (8), 1 (84), 3 (121), and 2 (132) are reproduced by permission from the *Revue Canadienne de Biologie* **24**, 243–276 (1965).

I extend my appreciation in particular to Dr. Pierre Bois, Professor and Director of the Department of Anatomy, Université de Montréal, for his support and for his valuable advice and to Dr. Serge Bencosme of Queen's University, Kingston, Ontario, who, through his enthusiasm for and devotion to the teaching of electron microscopy to students, was responsible for my interest in the field of ultrastructure.

I sincerely thank my assistants Miss Cécile Venne, Mrs. Andrée Aubin Delfour, and Miss Marie-Thérèse Fortin who investigated the inner ear in our laboratory. I also thank Drs. Réal Gagnon, Pierre Jean, and Bernard Messier for their valuable assistance and suggestions in the preparation of the original text.

I express my appreciation to Miss Emilienne Lambert for the drawing of the cell.

THE CELL

As seen in the light and in the electron microscope, cells display a marked variation of form as well as content.

Micrographs of the organelles, other inclusions, and membranes of the cell were taken from tissues at random. Variations will be seen in these organelles or inclusions in chapters covering the individual tissues.

The hepatic parenchymal cell has been chosen for a diagram since this complex polygonal shaped cell contains a greater variety of organelles and inclusions than most cells. It has a large rounded nucleus with moderately distinct concentrations of chromatin and one or more prominent nucleoli. The nuclear envelope is in the form of two membranes, an inner and an outer, which fuse periodically to allow a circular pore to appear. Ribosomes are present on the outer surface of the outer membrane and on sacs or tubules of the rough surfaced endoplasmic reticulum, with which it is continuous. The abundant rough surfaced endoplasmic reticulum is in continuity with a network of tubules of the smooth surfaced endoplasmic reticulum.

The Golgi apparatus or complex is a labyrinth of interconnected smooth surfaced sacs and tubules most probably in continuity with those of the endoplasmic reticulum. A stainable content is often seen in the peripheral sacs of the Golgi apparatus and in some of the endoplasmic reticulum.

Between the cisternae of the endoplasmic reticulum one sees a considerable number of mitochondria, organelles which are divided into two compartments by an inner membrane. This inner membrane folds to form flat or tubular crests or cristae which extend into the matrix of the inner compartment.

Other membrane-bound organelles, commonly seen in a section, are the *lysosomes* with densely stained content consisting largely of hydrolytic enzymes, *lipofuscin granules* which contain similar enzymes and pigments, *microbodies* or *peroxysomes* which contain another class of enzymes.

Fat droplets and numerous rosettes of densely stained glycogen granules are found near the mitochondria and between the cisternae or tubes of the endoplasmic reticulum.

Long slender cytoplasmic microtubules are seen near the cell surface and near the centrioles which, in turn, are usually located in the region of the Golgi apparatus.

In the liver, the hepatocytes are arranged in cords between sinusoids which transport venous blood from branches of the portal vein to the tributaries of the hepatic vein. During this circulation of blood from the intestine, the pancreas, and the spleen, through the liver to the right atrium of the heart, nutrients are absorbed by the hepatocytes and are resynthesized into utilizable forms for storage and for export. A part of the resynthesized or stored material is returned to the sinusoids as an internal secretion for use by other cells throughout the body, while bile is secreted into canaliculi for return to the small intestine. The bile canaliculi are formed by small segments of the surfaces of adjacent hepatocytes. These canaliculi lead to ducts which eventually empty the bile into the duodenum.

Slender cytoplasmic processes, microvilli, project into the lumen of the canaliculus. A number of fine filaments occupies the core of each microvillus and these extend for varying distances into the cytoplasm of the body of the cell.

The cell surfaces which face the blood sinusoids are also quite convoluted and slender processes project far into the space of Dissé which separates the extremely thin walled sinusoid from the hepatocyte.

Following the diagram throughout the book, the light micrographs have been designated by " L " at the beginning of their accompanying legends.

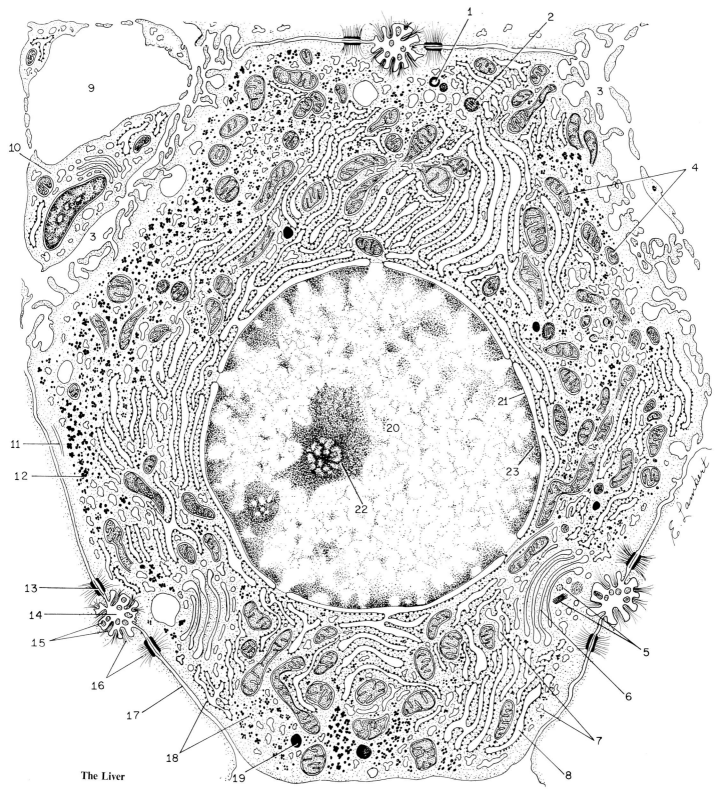

The Liver

1. lipofuscin granule
2. microbody (peroxysome)
3. extracellular space of Dissé
4. mitochondria
5. centrioles
6. Golgi apparatus or complex
7. smooth endoplasmic reticulum
8. granular endoplasmic reticulum

9. sinusoid
10. littoral reticuloendothelial cell
11. cytoplasmic microtubule
12. glycogen rosette
13. intercellular junctional complex
14. bile canaliculus
15. microvilli

16. filaments
17. plasma membrane
18. ribosomes
19. lysosome
20. nucleus
21. Inner nuclear membrane
22. nucleolus
23. chromatin

2   The Cell

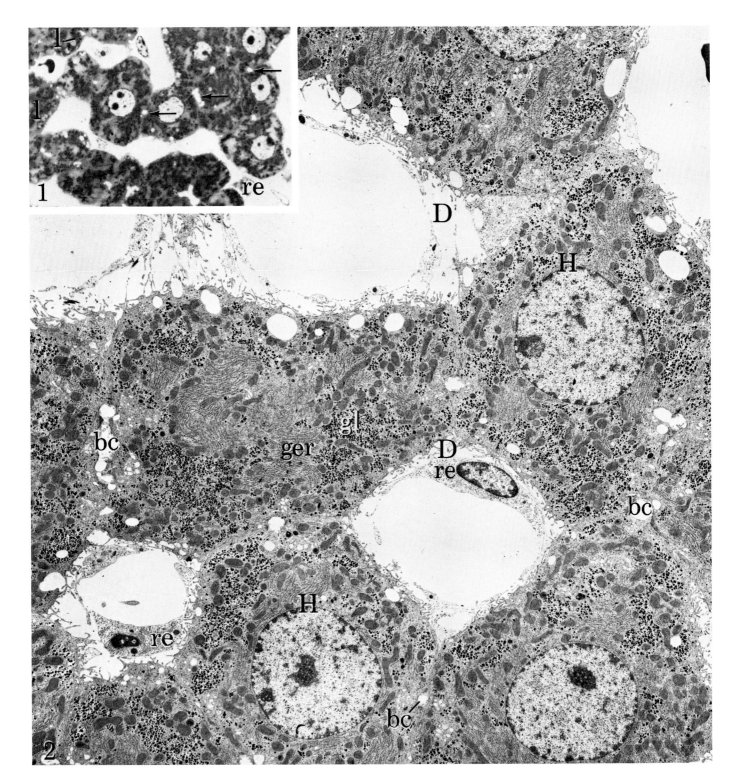

Liver Parenchymal and Sinusoidal Endothelial Cells

1. L. Cords of liver cells are bordered by sinusoids, in one of which a red blood cell remained after perfusion of the animal. Between the hepatocytes in the cord the small spaces with irregular borders (arrows) are the bile canaliculi. The large rounded nuclei often contain multiple nucleoli. In the cytoplasm the many homogeneous, bandlike structures are mitochondria. Other visible cytoplasmic inclusions are fine dense granules and pale circular lipid droplets (l). The ctyoplasm of the reticuloendothelial cells (re) which line the sinusoids appears as indistinct small masses some distance away from the hepatocyte. 600 ×

2. The elaborate fine structural detail of the hepatocyte and its relationship to the sinusoidal endothelial cell is indicated in this field. Cytoplasmic processes project from the hepatocyte (H) into the perisinusoidal space of Disse (D). The liver cell is notable for its abundance of granular endoplasmic reticulum (ger), accumulations of more densely stained glycogen granules (gl), and larger, dense lysosomal granules. The mitochondria appear as elongated or circular gray objects. A small part of the borders of neighboring hepatocytes outline the bile canaliculi (bc). 3000 ×

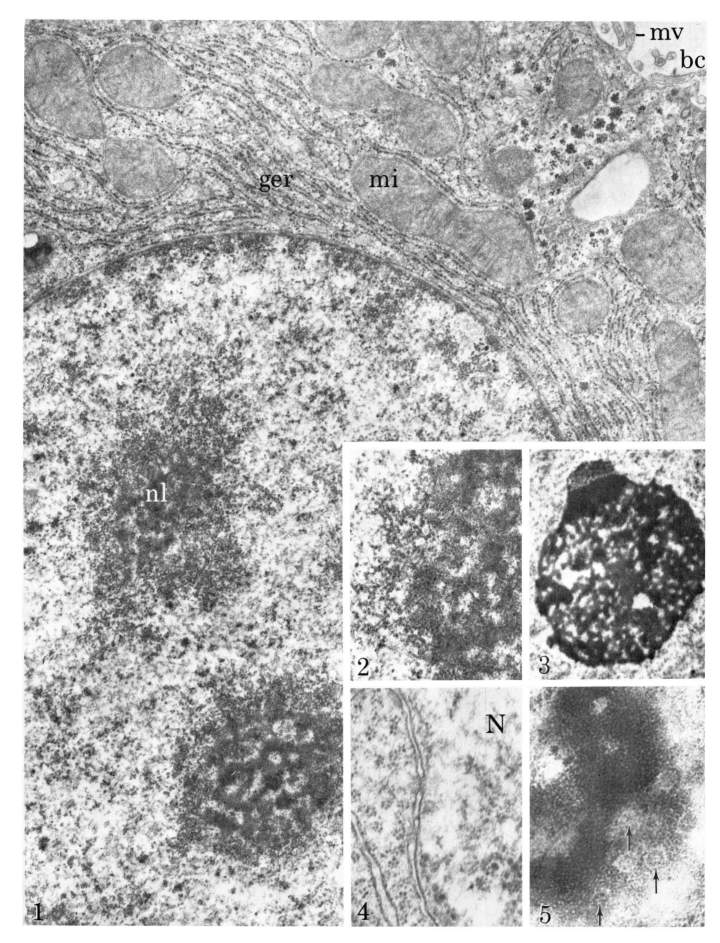

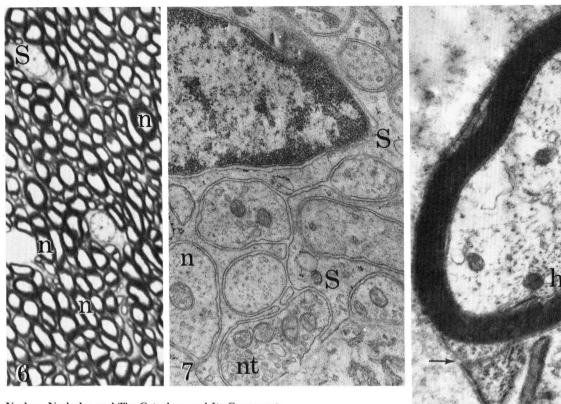

Nucleus, Nucleolus, and The Cytoplasm and Its Components

1. A part of a hepatocyte from the nucleus to the bile canaliculus. The nuclear chromatin stains as a fine granular material on a faint fibrillar background, but is more concentrated at the nuclear periphery and in a zone around the nucleolus. Darker and lighter chromatin granules are seen in different areas.

The three nucleoli (nl) are composed of much finer granules tightly packed into a network of dense bands with interwoven lighter zones. Some parts of the bands are more granular, others homogeneous. These bands produce the typical appearance of the nucleolonema, which had been described as a filamentous zone by light microscopists. The nucleolus is known to have a high content of ribonucleic acid (RNA).

The nucleus is surrounded by an inner and an outer nuclear membrane. The external surface of the latter is studded with ribonucleoprotein granules similar to those seen on the membrane of the granular endoplasmic reticulum (ger). The mitochondria (mi) are surrounded by an outer and an inner membrane, the latter of which folds inward to form crests. Toward the periphery of the cell the glycogen granules appear as rosettes of very dense particles. Microvilli (mv) project from the cell surface into the lumen of the bile canaliculus (bc). 26,000 ×

2. An enlargement of a portion of one nucleolus demonstrates the more and less granular zones of the nucleolonema. The more granular zones may represent nucleoprotein. 45,000×

3. The nucleolonema, the amorphous portion of the nucleolus, and a mass of perinucleolar chromatin are sometimes located in specific zones, as in this section of a large trigeminal nerve nucleus. 7700×

4. Purkinje cell—cerebellum. The outer and the inner nuclear membranes fuse periodically, and the resulting "pore" appears to be bridged by a thin diaphragm. 45,000×

5. Ameloblast. When seen in tangential section, nuclear pores are more numerous than anticipated from transverse sections of nuclear membranes. In a very thin section tangential to the nuclear surface the pore is outlined by a circle which represents the line of fusion of the membranes. 50,000 ×

Membranes—The Schwann Cell

Cells are surrounded by a plasma membrane. The myelin sheath of nerve fibers is particularly adaptable to the study of the "plasma" membranes. In polarized light studies a periodicity was detectable in each of the dense myelin sheaths which surround the axons. The Schwann cell has long been considered as the source of the myelin sheath.

6. L. Trigeminal nerve-myelinated nerve fibers. The Schwann cells (S) are found among the many myelinated nerve fibers.(n). The myelin appears as dark rings around the pale axons. 800×

7. Schwann cell and unmyelinated autonomic nerve fibers of the intestine. Nerve fibers (n) are embedded in deep folds of Schwann cell cytoplasm (S). Even the smallest of nerve fibers carry their neuronal plasma membrane with them, and a space remains between them and the external surface of the invaginated Schwann cell plasma membrane. Even the nerve terminals (nt), in which many vesicles can be detected, retain a surface plasma membrane distinct from that of the Schwann cell. 20,000×

8. The myelin sheath is formed by multiple layers of Schwann cell plasma membrane wrapped around the nerve fiber. The point of the infolding of the surface of the Schwann cell (arrow) and the head of the fold (h) can be seen. This head, wherein the Schwann cell membrane folds back upon itself, is immediately exterior to the axonal plasma membrane. The cytoplasm of the Schwann cell envelops the axon. The latter remains extracellular in relation to the Schwann cell. The Schwann cell cytoplasm contains the organelles common to most cells. The axon has a limited number of types of organelles. The larger organelles in the axon are mitochondria. The very small circles are cytoplasmic microtubules sectioned transversely. 30,000×

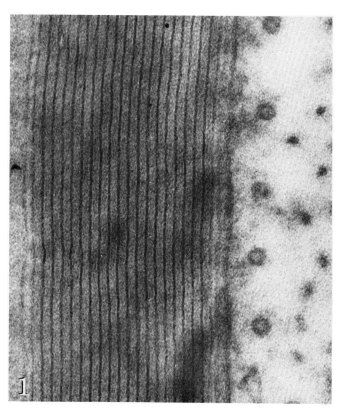

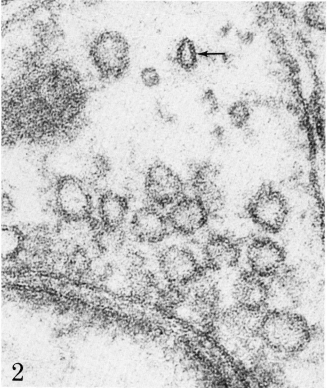

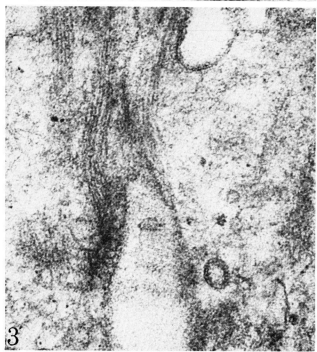

Membranes

Early quantitative studies were done on red cell ghosts in order to determine the lipid content of cell membranes. From these studies the earliest hypothetical model of the molecular arrangement of lipids in membranes was proposed. The periodicity of the myelin sheath, produced by layers of Schwann cell plasma membrane, has made it particularly interesting for the study of membrane substructure.

1. Peripheral nerve—trigeminal nerve. Each membrane has a lighter central layer between inner and outer dense leaflets or laminae. A periodicity of the myelin sheath reflects the repeated layers of the plasma membrane of the Schwann cell as it makes successive turns around the nerve fiber. The major dense lines or periods result from the fusion of the inner dense laminae of the plasma membranes as each turn contacts the preceding layer. These major lines are separated by approximately 120 Å. The outer laminae of membranes, which come into contact when the Schwann cell membrane is first infolded, form a less dense "interperiod line." Thus each period represents two complete membranes. The interperiod line can be seen as two separate, thin, moderately dense lines in some instances.

The axoplasm is bounded by a distinct trilaminar membrane of dense, light, and dense layers. This membrane is separated from the myelin sheath by a space of approximately 150 Å. The axon contains circular profiles which are cytoplasmic microtubules in transverse section, an irregular tubule of the endoplasmic reticulum, and transversely sectioned filaments. The lumen of the cytoplasmic microtubule is approximately 100 Å in diameter and the wall 60 Å in thickness. The filaments are approximately 50 Å in thickness. 240,000×

2. Cerebellum. The three layers have been demonstrated in all types of membranes in the cell, but this feature is more readily seen in some membranes than in others. Recently, a globular-appearing substructure has been demonstrated in some internal membranes. A vesicle or transversely sectioned tubule in an axonal synaptic ending in the cerebellum demonstrates this globular appearance (arrow). Other vesicles and the plasma membranes of the synapse are mostly trilaminar. In oblique and surface views of the walls of vesicles there is evidence of fine lines in parallel. 205,000×

3. Apical vesicles in cells of the intestinal epithelium which is actively engaged in absorption are often considerably dilated. A surface view of one of these dilated vesicles presents an appearance of lines in parallel. Filaments from the rootlets of the microvilli are in close association with the wall of the large vesicle. 128,000×

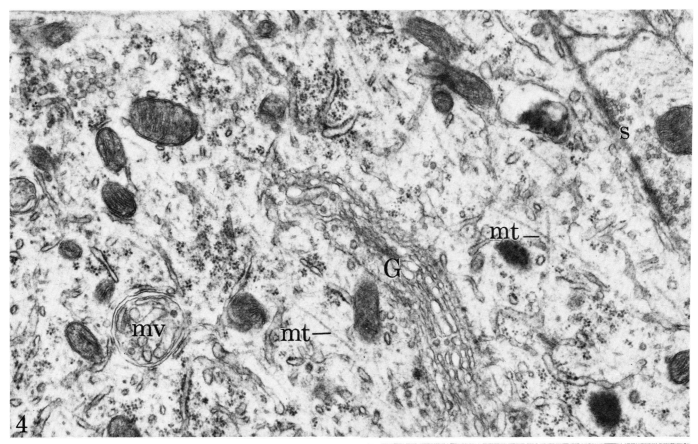

The Golgi Complex or Apparatus and Endoplasmic Reticulum

4. The Golgi complex (G) is most commonly located in a juxtanuclear position, but in the neuron it is more widely distributed through the cytoplasm. This example is from near the periphery of the perikaryon as evidenced by the axosomatic synapse (s). In the central nervous system the Golgi complex extends out into the larger dendritic processes of neurons. The Golgi complex appears as several layers of interrupted sacs, stacked one upon the other and generally associated with more widely spread vesicles at its periphery. Commonly associated with the Golgi complex are dense granules, cytoplasmic microtubules (mt), multivesicular bodies (mv), and elements of the agranular reticulum. The microtubules are long, slender organelles with an outside diameter of about 210–230 Å and are distributed throughout the cytoplasm of all types of cells. The function of the multivesicular body is not clearly understood. The agranular endoplasmic reticulum appears as a branching network of tubes in some instances. Mitochondria, some with transversely and some with longitudinally oriented crests, are scattered throughout the cytoplasm. The numerous granules in the cytoplasm and on membranes are ribosomes which are known to be responsible for the basophilic stain in light microscopy. It is on the ribosome particles that proteins are synthesized. They occur singly or in clusters called polysomes. In addition they are located on the outer surface of the outer nuclear membrane and of the membranes of the long tubes or cisternae of the granular endoplasmic reticulum. Larger accumulations of ribosomes and of the granular endoplasmic reticulum have been termed Nissl bodies in the neuron. These appeared as localized areas of intense basophilia in the light microscope. Other membranes, some continuous with the membranes of the granular endoplasmic reticulum, are elements of the smooth-surfaced or agranular reticulum. 30,000×

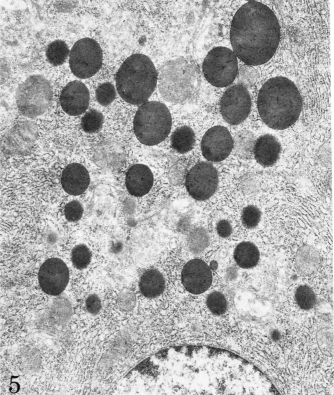

The Pancreatic Acinar Cell

5. Both moderately dense and dense granules develop in the Golgi region. The larger granules are generally seen at a greater distance from the Golgi complex. 14,000×

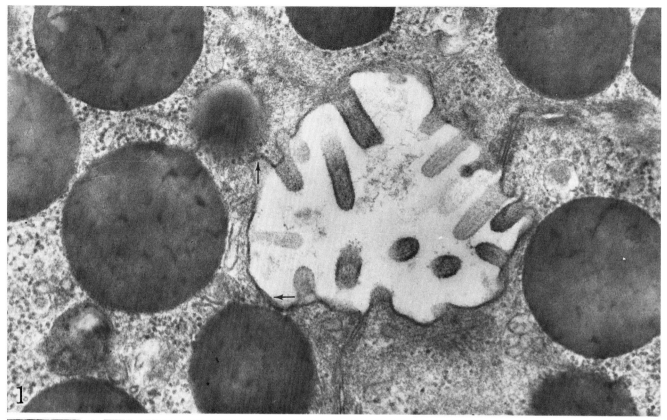

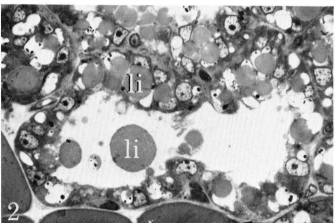

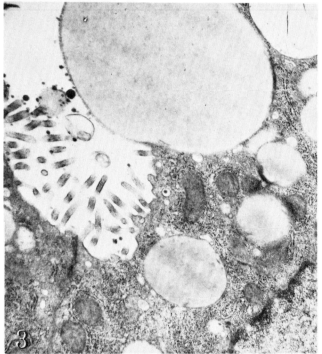

Secretion—The Pancreatic Acinar Cell

1. Apexes of acinar cells, interconnected by tight junctions, form the borders of the lumen of the acinus. Microvilli with filamentous cores project into the lumen. Large, dense zymogen granules occupy much of the apexes of the cells. One of these granules appears partially emptied. Filamentous or microtubular stems associated with these granules appear to reach and to blend into the plasma membrane (arrows). 55,000 ×

Lipid Secretion, Fat Cells—Mammary Gland

2. L. The alveolar cells of the mammary gland secrete fats in large quantity into the lumen. The cytoplasm of many cells appears almost completely occupied by large homogeneous lipid droplets (li). Other large clear intracellular spaces contain dense granules. These are accumulations of milk proteins, which are also secreted into the lumen of the alveolus. The alveoli are surrounded by fat cells, which appear as large

accumulation of lipid with a very thin cytoplasmic border. The nucleus is small and elongated (arrow). 500 ×

3. The lipid droplets protrude far into the lumen before they are released through the plasma membrane. The smaller dense granules in the lumen are the milk protein. The granular endoplasmic reticulum appears as parallel, moderately closely applied layers of ribosome-studded membranes. Osmic acid fixation-ethanol dehydration embedded in Epon. 16,500 ×

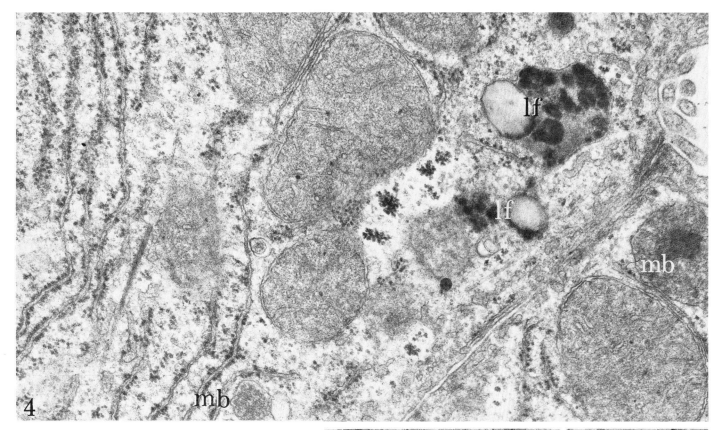

Mitochondria, Endoplasmic Reticulum, and Granules

4. In the hepatocyte the endoplasmic reticulum appears as narrow channels with the granular and the agranular types in continuity. The latter is in close contact with the mitochondria. The crests, or cristae mitochondriales, of the inner membrane are closed at their apex in some cases. In others, these crests appear isolated from the inner circumferential membrane and in still others, as complete transverse bridges. These effects may all be produced by the direction of section through a fold, but complete membranous bridges across the mitochondrial matrix do occur. Dense granules are present in most of these mitochondria. Calcium and other divalent cations are bound to these sites. The size of the granule can be modified experimentally by varying the availability of these ions to cells or mitochondria in experimental media. The large granule with both light and dark zones is a lipofuscin granule (lf). These granules increase in number with age. The dark pigment portion is believed to represent insoluble cellular debris. An active functional role has been suspected for these granules since esterase activity has been localized within them. Another granule, surrounded by a single membrane, with a more uniform granular background and a darker paracrystallin central body, is a microbody (mb). Recently, a high level of uricase activity and peroxidase have been found in microbodies. An alternate name "the peroxysome" is commonly applied to this organelle.

The clusters of granules, larger and more dense than the ribosomes, are glycogen rosettes. 55,000 ×

5. A large lysosome in a follicular cell of the thyroid gland of an adult rat is surrounded by a three-layered membrane. The membrane and the dense content are separated by a space of approximately 75 Å. There are suggestions (arrows) that another membrane may be in contact with the dense material, but it is almost completely obscured by the content. 75,000 ×

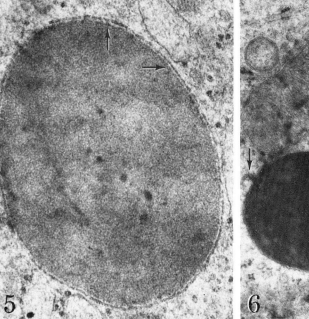

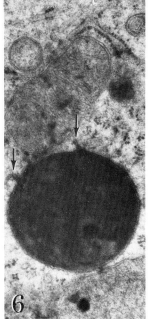

6. Lysosomes in the epithelial cells appear to be surrounded by two membranes. Microtubular stems with dense content project from the surfaces of these lysosomes (arrows), kidney. 55,000 ×

The hydrolytic enzymes located in the lysosomes include acid ribonuclease, acid deoxyribonuclease, acid phosphatase, phosphoroprotein phosphatase, cathepsin, collagenase, α-glucosidase, β-N-acetylglucosaminidase, β-glucuronidase, β-galactosidase, α-mannosidase, aryl-sulfatase, and hyaluronidase.

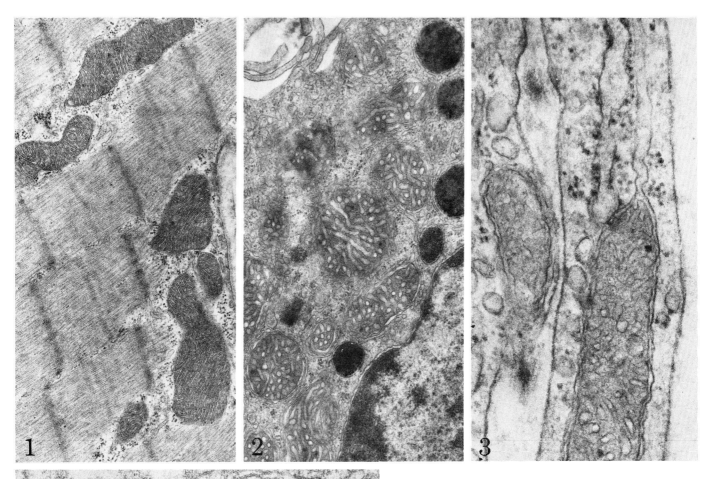

Mitochondria

The enzymes involved in oxidative phosphorylation, coupled with respiration, are located in mitochondria. The reactions taking place within this organelle are the source of energy which has led to the mitochondrion being termed the "powerhouse of the cell". The mitochondria provide the energy for lipid metabolism as well as the adenosine triphosphate required for muscle contraction. They have also been found to contain contractile proteins and a small quantity of deoxyribonucleic acid. In time lapse cinematographic studies on living cells mitochondria appear to be quite motile organelles.

Methods are constantly being developed for the localization of sites of synthesis of various compounds and of sites of enzymatic activity within the cell. These methods have included fluorescence microscopy, histochemistry, cytochemistry, and autoradiography. All of these methods have contributed considerably to our understanding of the functional significance of the various components of the cells and tissues. A few examples of the results obtainable by some of these methods have been included. It has been possible by electron microscopy to locate enzymatic reactions on specific sites on the plasma membrane and on various organelles. The results are obtained by the liberation, at the site of the reaction, of radicals which are then precipitated with heavy metals. The precipitate can be found in the sections by light and electron microscopy.

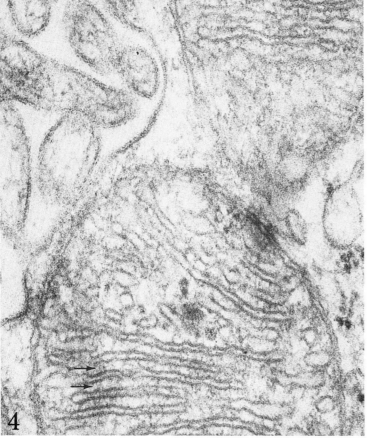

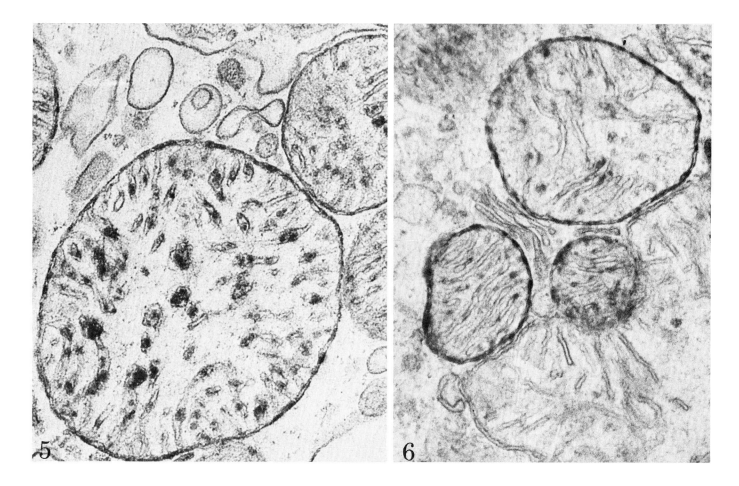

Mitochondria

1. Cardiac muscle. In aldehyde and osmium-fixed tissue the mitochondrial matrix is quite dense in most cells, while the intracristal space is quite clear. Mitochondria are often very intimately associated with lipid droplets. 20,000×

2. In the adrenal cortex the mitochondria contain closely packed tubular crests which, when sectioned transversely, appear as numerous small circular membranous profiles within the organelle. The tubular crests are sectioned both transversely and longitudinally in mitochondria of this cell in the reticularis. Granules with a clear zone intervening between a central density and a membrane are frequently encountered. 25,000×

3. Mesothelial cell—tunica vaginalis testis. Membranes of an irregular tube of the endoplasmic reticulum are in continuity with the outer mitochondrial membrane. 70,000×

4. Gastric gland. The matrix of the very large mitochondria in the parietal cells does not stain densely after aldehyde and osmium tetroxyde fixation. The mitochondrial membranes appear trilaminar or globular. Filaments are located between the plasma membrane of an intracellular canaliculus and a mitochondrion. An internal component of the mitochondrial matrix appears filamentous (arrows). 112,000×

5. Reticular cell, thymus, rat. An example of the cytochemical localization of succinate dehydrogenase reaction product on mitochondria. The precipitate has been localized to the space between the outer and the inner membranes and between the membranes of the crests. Some tubular crests appear to be filled with the reaction product. (Copper ferrocyanide method.) 56,000×

6. The site of the reaction product of cytochrome C reductase activity has been localized between the outer and the inner mitochondrial membranes and, to a lesser extent, in the cristae. Some of the mitochondria in this field show minimal amounts of reaction. (Copper ferrocyanide method.) 58,000×

Figures 5 and 6 contributed by Dr. Takashi Makita (unpublished).

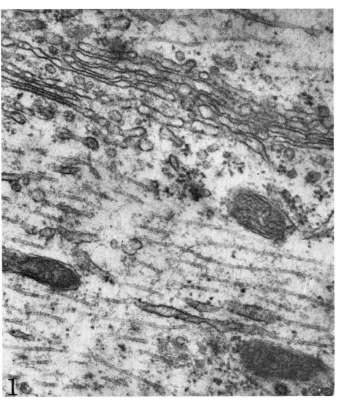

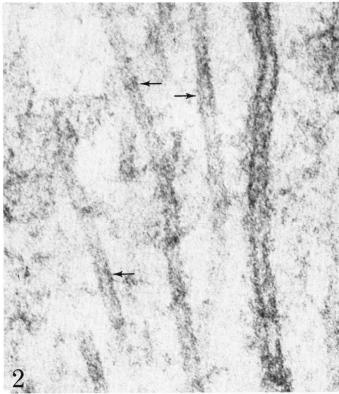

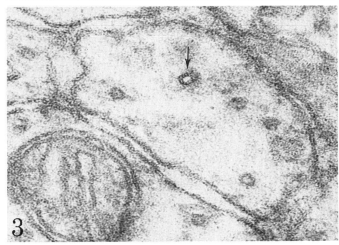

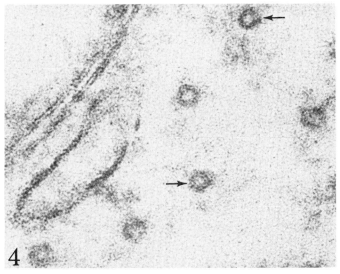

Cytoplasmic Microtubules

The earliest demonstrations of these organelles in neurons were in dendrites of osmium tetroxide-fixed tissues. It was believed that these microtubules were primarily associated with the dendritic process. With the advent of fixatives which penetrate the perikaryon and myelinated axons more readily it has been established that they extend throughout the cytoplasm of the neuron. Cytoplasmic microtubules have been found to be constitutents of all types of cells.

1. Cerebellum—Purkinje cell. Golgi membranes, both the granular and the agranular endoplasmic reticulum, the mitochondria, and the cytoplasmic microtubules extend into the larger dendrites of neurons. The smooth-surfaced endoplasmic reticulum appears as irregular tubes which are either very closely associated with or in continuity with the cytoplasmic microtubules. 46,000×

2. In the follicular cells of the cricket ovary there are many cytoplasmic microtubules near the cell borders. When the sections are sufficiently thin, three layers (arrows) are demonstrable in the microtubular wall. In thicker sections, the central light layer of approximately 20 Å within a sharply curved wall would be masked by an overlapping of the dense leaflets. 240,000 ×

3. Unmyelinated axons—cerebellum. Three layers, more or less distinct, can be seen in the plasma membrane. A less distinct trilaminar appearance is present in the membranes of the mitochondria and of the endoplasmic reticulum. In most transversely sectioned cytoplasmic microtubules the substructural appearance of the wall is indistinct, but in one (arrow) a globular appearance is present. 145,000×

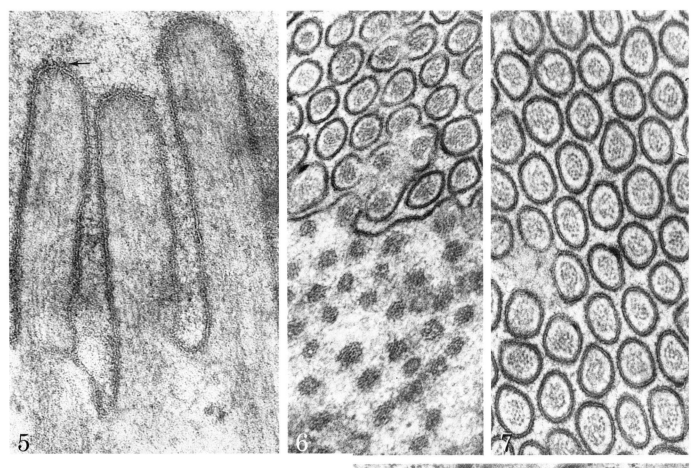

5

6

7

4. Unmyelinated axons—cerebellum. Other examples of the trilaminar structure (arrows) are seen in the walls of microtubules and of the endoplasmic reticulum and in the plasma membranes. In membranes which are not perfectly oriented the three layers are not detectable. 240,000 ×

Microvilli, Filaments, and Microtubules—Duodenal Crypt Epithelial Cells

5. At a high magnification the outer layer of the plasma membrane of microvilli appears to be carried outward (arrow) to form the filamentous extraneous coat. This has been termed a glycocalyx and appears to be a protective coat elaborated by the cell membrane. The filaments of the core of the microvilli appear to contact the inner aspect of the plasma membrane. The outer and the inner leaflets of the plasma membrane appear equally dense. 170,000×

6. Sections which transect the microvillous border and pass into the cell apex proper demonstrate the microvilli and their filamentous rootlets in transverse section. 76,000×

7. In a section which passed transversely through the microvilli, a moderately clear space is found between the filamentous core and the plasma membrane. The filaments range in number from twenty to thirty per microvillus. 108,000×

8. Deeper within the cell, filaments, often very densely stained, form branching bundles. These were termed tonofibrils in light microscopy. Cytoplasmic microtubules are often closely associated with these bundles. Vesicles with coated surfaces and a dense content are seen nearby. 50,000×

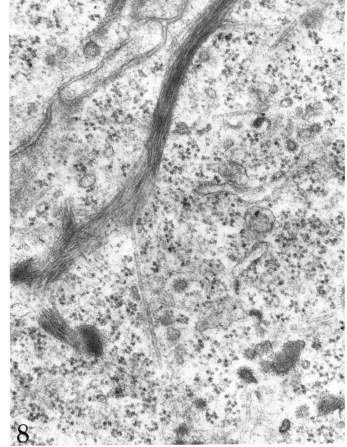

8

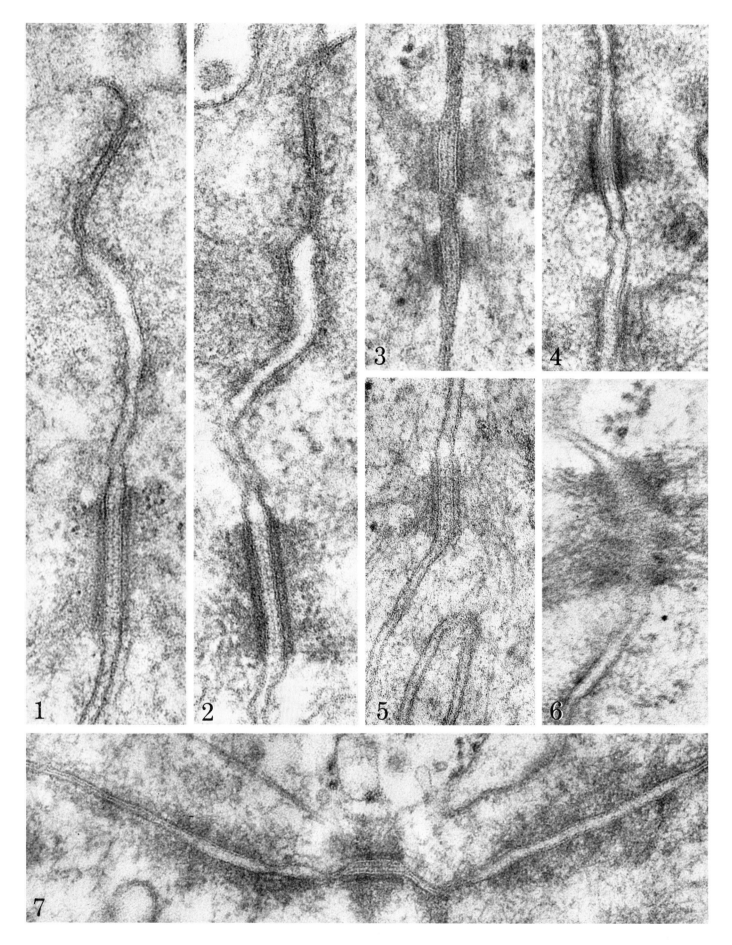

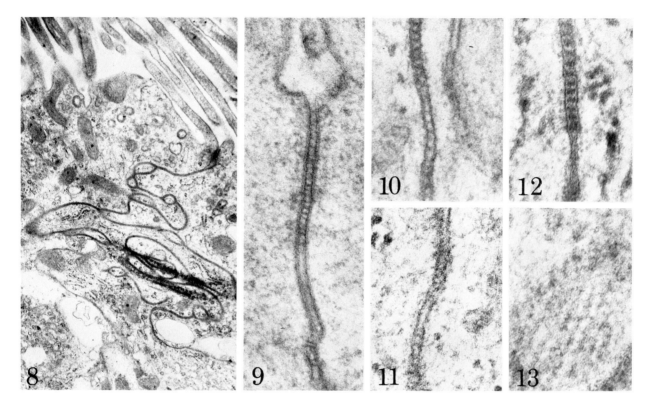

Intercellular Junctional Complexes—Duodenal Epithelium

1. The ultrastructural appearance of the contacts between cells varies with their location. The junctions between the apices of adjacent epithelial cells demonstrate some of these modifications in close succession. The terminal bar, if the term can be applied to the ultrastructure of this region, has been subdivided into: (a) the zone of close contact between the apical portions of adjacent cells termed the zonula occludens; (b) the zone wherein the intercellular space increases, the zonula adherens; and (c) the localized areas with associated filaments and dense intracellular plaques, long known as desmosomes. This type of junction has also been termed the macula adherens. The desmosomes are multiple specialized zones of contact which occur repeatedly between the cells in addition to participating in the terminal bar. 114,000×

2. In most instances, the central line in the macula adherens or desmosome is indistinct. Fine lines bridge the gap between this central line and the adjacent membranes. A similar network of bridges spans the space in the zonula occludens and in the zonula adherens. A slight amount of obliquity of section results in a loss of the interspace between membranes in localized zones. 155,000×

3. The parallel lines are the dominant feature in the desmosomes.
A network of filaments, in some instances appearing to arise from the membranes project into the cytoplasm and into the intercellular space. 108,000×

4. One desmosome has a clearly defined central line, while the other desmosome appears to contain a series of curved intercellular filamentous bridges. Other bridges occur outside of the desmosomes. 100,000×

5. In this desmosome there appears to be a crossed network of curved filaments within the intermembranous space. 120,000×

6. In an oblique section the desmosome appears as a mass of lines in varying directions, but mainly in a cell to cell direction.

It is possible that this impression could be created by the overlapping of intracytoplasmic filaments from the two cells, but a crossover of filaments is a distinct possibility. 152,000×

7. In a horizontal section through the apices of adjacent cells the terminal bar can be followed from a zonula occludens through the zonula adherens to the macula adherens and back through the adherens to the level of the zonula occludens at a distance from the point of origin. In this case there appears to have been a fusion of the outer layers of adjacent membranes in both segments of the zonula occludens. 127,000×

Septate Desmosomes

FIGS. 8, 9, 10, 11, 12, AND 13. FOLLICULAR EPITHELIUM OF THE CRICKET OVARY

8. Insect epithelia often have a very tortuous area of contact between the apical portions of adjacent cells. 13,000×

9. Intestinal epithelium—*Drosophilia melanogaster* third instar larva. The transverse bars of the septate desmosome often reach the cell apex and reappear at close intervals throughout the supranuclear portion of the cell. 129,000×

10. On occasion a cruciate network is formed by the crossing of bars in different directions. Some of these intercellular bridges have a trilaminar appearance. 136,000×

11. Some of the bars are distinctly curved and exhibit a central pale lumen. 136,000×

12. In other views the transverse bars appear to be made up of obliquely sectioned rod like subunits. 136,000×

13. In a surface view of the desmosome a series of irregular lines in parallel can be seen. 160,000×

These findings have been interpreted differently by various authors, each being influenced by his acceptance of various membrane concepts. The author and co-workers have interpreted the various features demonstrated in these micrographs as evidence of a crossover of a filamentous membrane substructure between cells in this type of desmosome.

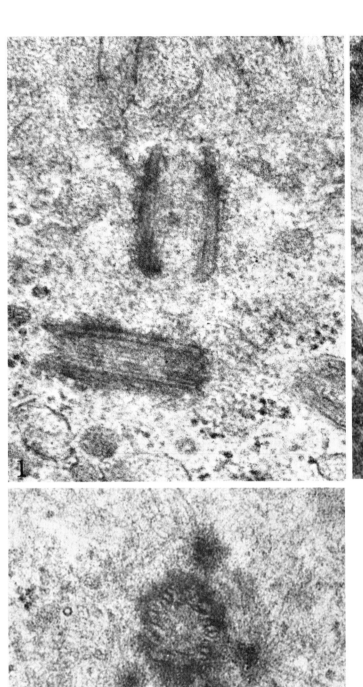

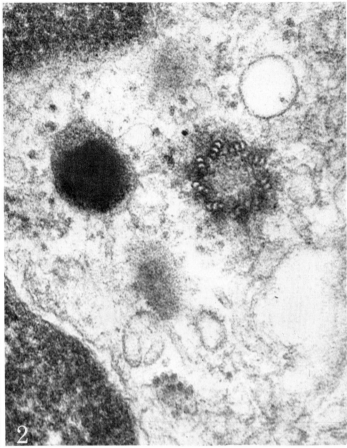

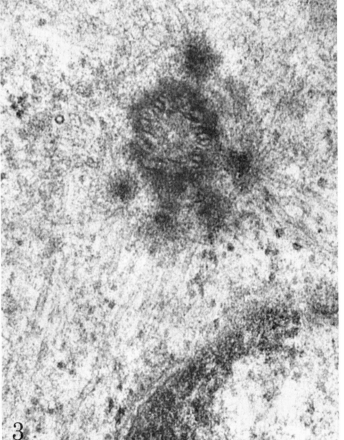

The Centriole

Two or more centrioles are found in cells which retain the potential to divide. In most cells the centrosome is found in the juxtanuclear Golgi region. In epithelial cells the pair of centrioles is often located in the apex.

1. In the thyroid follicular cell two apical centrioles, both sectioned longitudinally, are oriented at right angles to one another. A series of parallel lines is visible in the long axis in each centriole. Filaments extend beyond the end of one centriole, while a group of filaments extends up into the microvillous border from the side of the other. 94,000 ×

2. Reticuloendothelial cell of a liver sinusoid. A transverse section of a centriole in the Golgi zone demonstrates that the parallel longitudinally oriented structures are twenty-seven tubes arranged in nine triplets. Each of the microtubules within the triplets shares its wall with the adjacent tubule. The membrane of a dense granule in the immediate vicinity appears to be continuous with a microtubular stem. 90,000 ×

3. In a reticuloendothelial cell of a liver sinusoid the section has passed through several centriolar satellites, each of which appears to be composed of a mass of radiating filaments. Each is attached to the centriole by filaments and both microtubules and filaments radiate out from the satellites. The cytoplasmic microtubules appear to take form among the filaments of the satellites. A cytoplasmic microtubule is sectioned transversely. 100,000 ×

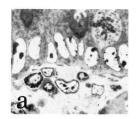 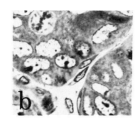 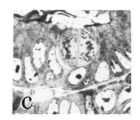 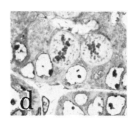

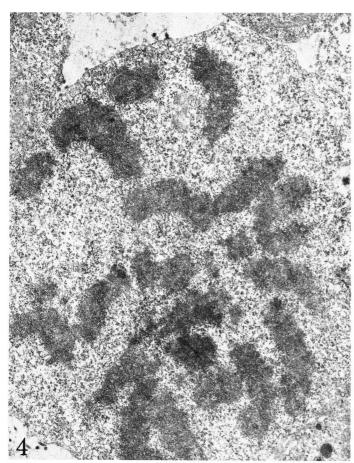 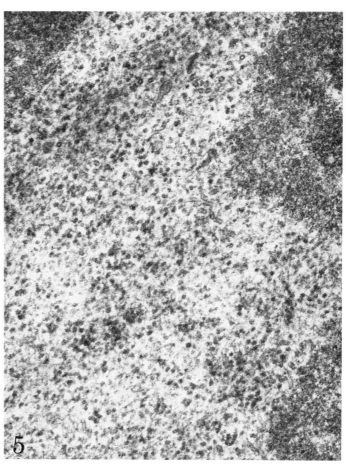

Mitosis—Duodenal Epithelium

Shortly after duplication in late interphase, the two new pairs of centrioles migrate to opposite poles of the cell. The nuclear membranes disintegrate and the nucleoleus disappears as the distinctive changes of the four phases of mitosis begin to appear.

Insets (L):
a. The large, pale cell which is completely elevated above the nuclear level of the neighboring cells is in late prophase. The chromatin has become arranged into elongated masses typical of the chromosomes. In late prophase they migrate toward an equatorial plane of the cell. 520 ×

b. Cells in the metaphase wherein the chromosomes are aligned along the equator. The clear zone bordered by a few mitochondria is the space occupied by the mitotic spindle. 520 ×

c. A cell in anaphase wherein one-half of the new chromosomes migrate toward one pole and the other half to the opposite pole of the cell. A peripheral cleavage furrow appears as the new membrane separating these cells is formed. 520 ×

d. A cell in telophase, during which only a cytoplasmic bridge connects the two daughter cells. In each of the daughter cells the reorganization of a distinct nucleus and definitive organelles follows rapidly. 520 ×

4. Duodenal crypt epithelial cell. The cell in prophase at low magnification in electron microscopy contains dense, finely granular chromosomes against a lighter background of more discrete ribosomal granules. A few organelles appear at the periphery. 14,000 ×

5. At higher magnification cytoplasmic microtubules, some sectioned transversely, others obliquely, and filaments can be discerned among the ribosomes. 66,000 ×

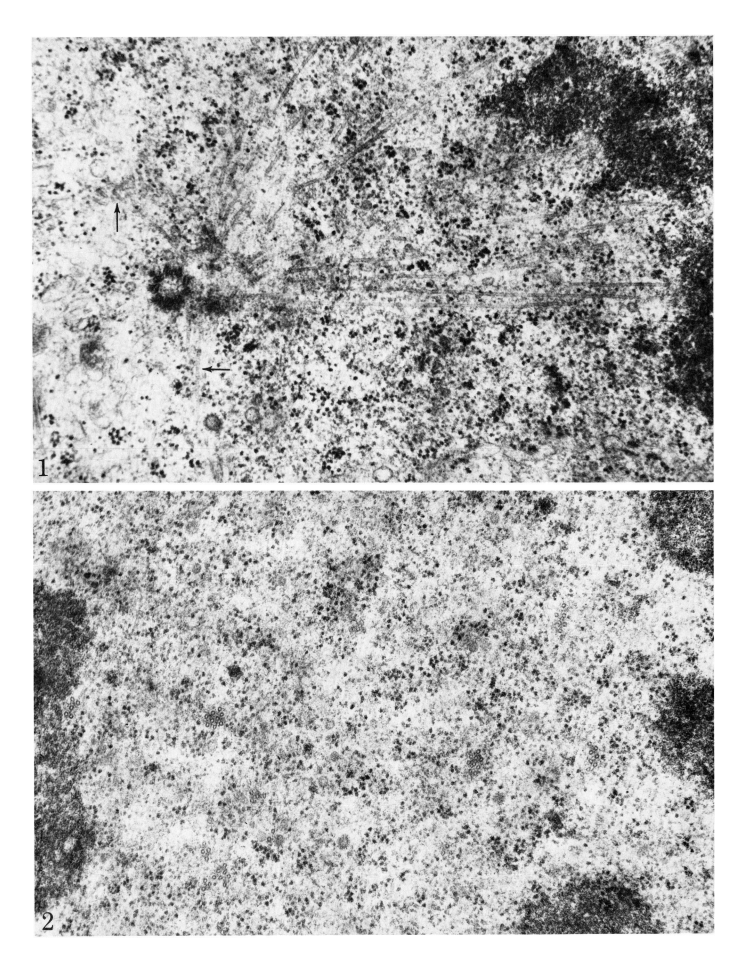

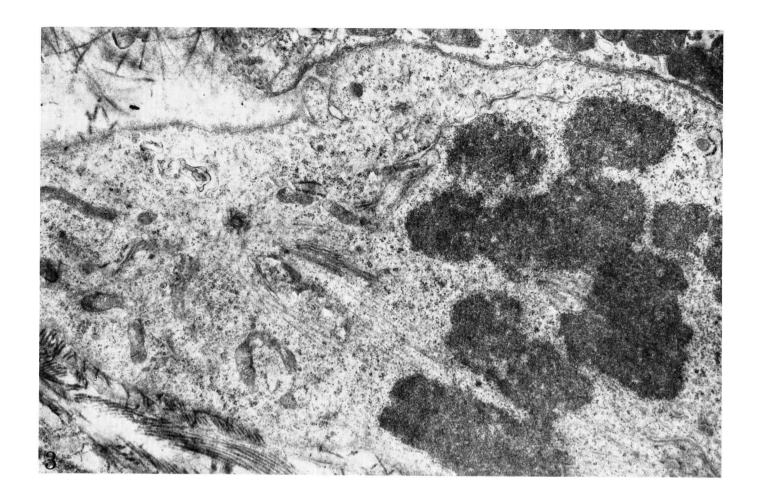

Mitosis—Metaphase

A birefringence within the spindle and the asters of living metaphase cells, when examined with polarized light, was interpreted as being due to spindle and astral fibers. Studies in the electron microscope have shown that the birefringence was produced by groups of hundreds of microtubules arranged more or less in parallel. The majority of these interconnect the chromosomes which have become arranged in pairs along the equator to their centrioles. Others pass through between the chromosomes as continuous microtubules to the centrioles at the opposite pole of the spindle.

1. Cytoplasmic microtubules radiate from the centriolar satellites to the chromosome. In a narrow zone near the chromosomes the microtubules give way to a much larger number of indistinct filaments in the kinetochore. Obliquely sectioned astral microtubules and filaments (arrows) radiate out in a much wider arc toward the periphery. 70,000 ×

2. A transverse section through a very small part of a spindle near a chromosome includes many microtubules in groups of varying numbers. Groups of nine are quite common among these. 60,000 ×

Mitosis—Telophase

3. In a fibroblast in telophase the Golgi apparatus begins to reappear as elongated parallel tubes in the spindle. One segment of the nuclear membranes has reappeared. Ribosomes have made their reappearance on the outer nuclear membrane. The microtubules of the direct fibers can be seen beyond the chromosomes passing toward the midbody. The membranes of other organelles are making their appearance in the equatorial zone. 17,500 ×

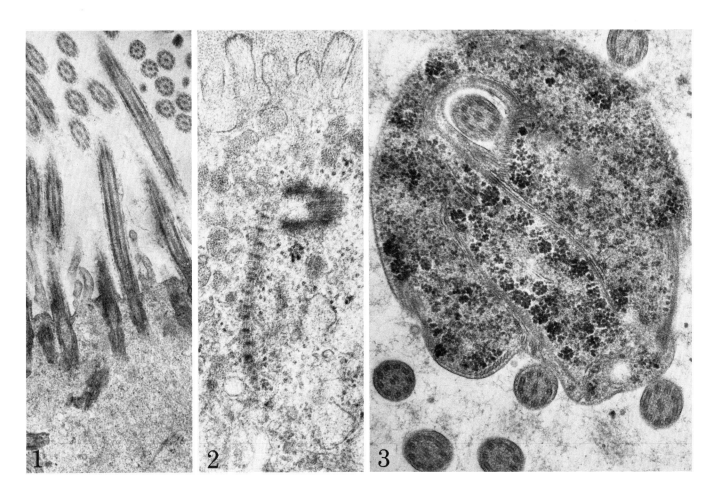

Cilia and Flagella

Motile cytoplasmic processes are found on the surfaces of protozoa. On some types of cells in multicellular animals cilia are numerous, but quite commonly a single cilium graces the cell surface.

1. Many cilia distinguish this follicular type of cell from others in the pars distalis of the hypophysis. The axoneme of the cilia as seen in longitudinal and transverse section, contains an internal axial substructure of nine peripheral doublets and two central microtubules and is surrounded by a plasma membrane. The basal bodies of the cilia are situated within the apex of the cell and resemble the centriole from which they were derived. In the differentiation of these cells hundreds of ciliary basal bodies develop by repeated divisions. This is followed by the formation of the ciliary processes. 60,000 ×

2. A periodicity is present in the rootlet of a basal body of a single cilium in a thyroid follicular cell. Microtubules radiate out from the rootlet of the cilium. 60,000 ×

3. Protozoa are normal inhabitants of the intestine of many animals. This single-celled animal, a motile inhabitant of the duodenal crypts in the laboratory rat, has six peripheral flagella and one within its buccal stoma or cytostome. The flagella are employed in a whiplike fashion in the locomotion of these animals. The flagellum like the cilium has a membranous sheath and an axoneme which is composed of nine peripheral doublets and two central microtubules. There are numerous peristomal and peripheral microtubules; microtubules are numerous as well among the many ribosomes in the cytoplasm. 45,000 ×

4. Many ciliated cells are present in the tracheal epithelium. Numerous long, thin microvilli are found between the cilia. An irregular filamentous tuft appears at the apex of each microvillus. A series of fine spikelike projections are demonstrable along the surfaces of the peripheral doublets as well as on the central microtubules of the cilia. In some areas these appear to bridge the gap between the peripheral and the central doublets. The plasma membrane forms a rather irregular loose sheath around the ciliary process. 60,000 ×

5. An enlargement of the shaft of a cilium demonstrates the details of these doublets, axial microtubules, and their lateral projections. 105,000 ×

6. In transverse section it can be seen that the microtubular subfibrils of the doublet share the same wall. An incomplete microtubule "B" appears to be attached to a completely circular "A" microtubule. Hooklike processes extend out from the border of the "B" tubule, and faint spokes radiate out from the central microtubules toward the doublets. Central microtubules are separated by a variable distance of from 30 to 150 Å. 68,000 ×

The direction of the rapid beating movement of cilia is at right angles to a line joining these central microtubules. The direction of this line is said to be constant in all cilia of a given cell, although at a distance from the apex of the cell this does not always hold true.

7. In the end piece of a sperm tail hooklike processes extend from the "B" tubule toward the "A" tubule of the next doublet. Faintly stained fibrillar material similar to that of the spokes which radiate out from the central microtubules partially envelops some of the peripheral doublets. 140,000 ×

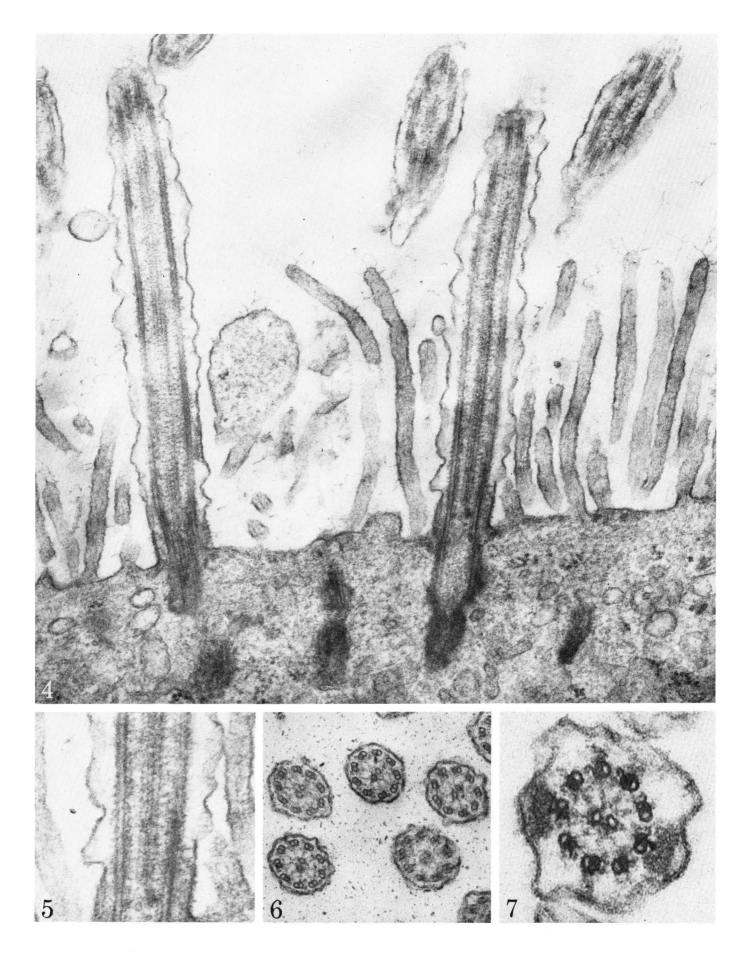

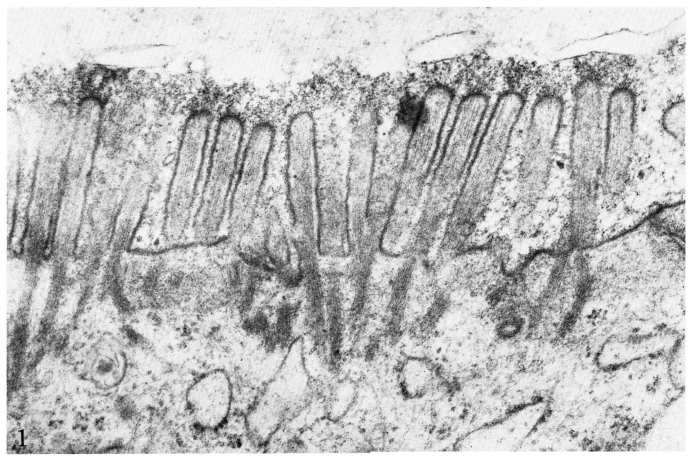

External Coats of Cell Membranes

1. The mucopolysaccharide tufts which have been reported on the apices of intestinal microvilli vary remarkably in different species. In the dog intestine a considerable layer is created by these filamentous extensions, which form a protective mat across the apices of the microvilli. As in the rat intestine, it appears that these extend out from the plasma membrane of the microvilli. It is believed that these coats are elaborated by the cells whose surfaces they cover. The layer has a high content of carbohydrates in the form of polysaccharides. An acid mucopolysaccharide content in the extraneous coat of many cell membranes, including those of the intestinal epithelium, has been demonstrated by a specific ruthenium red stain. Extraneous coats differ in appearance and in thickness and can be altered experimentally. 68,000 ×

2. By freeze etching a block of duodenal epithelium, which had been immersed in a 15% dimethylsulfoxide (DMSO) solution after a 5-minute prefixation in a glutaraldehyde-DMSO solution, a surface view has been obtained of the microvilli and of some organelles. The surfaces of the microvilli are quite irregular, suggesting a filamentous coat. In the cell apex many transversely fractured filaments project above the background cytoplasm. Longitudinal views of the numerous filaments in the cell apex have been difficult to obtain by this method. A winding tubule (arrows) can be followed from the apical surface to the level of horizontal tubes deeper within the cytoplasm. A portion of the etched membrane of this tube appears to be covered by a less well-developed filamentous coat than that which covers the microvilli. 90,000 ×

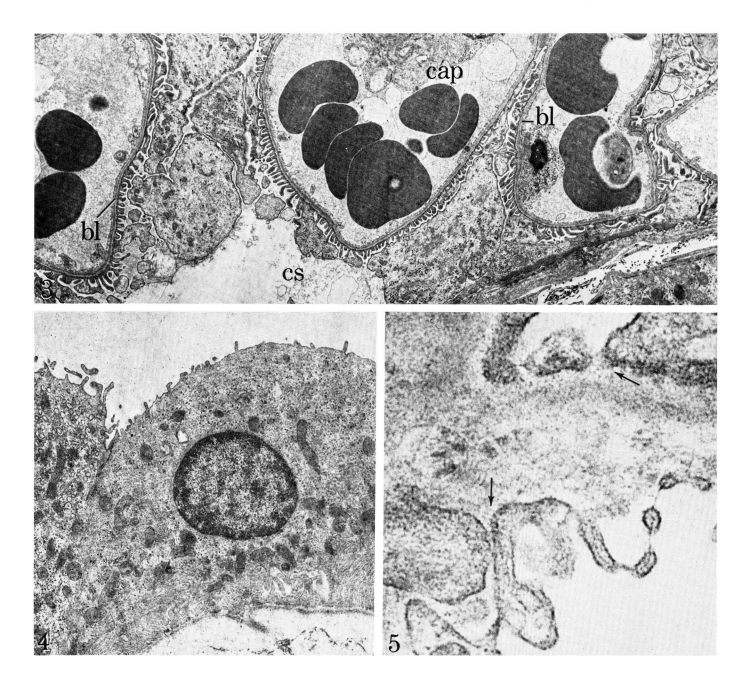

External Laminae

Basal laminae most commonly appear as layers of homogeneous material a few hundred angstrom units in thickness and separated from the base of epithelial or endothelial cells by a narrow lighter band. In other instances, such as in the cases of muscle or Schwann cells, a circumferential lamina surrounds the cell completely. This is a mucopolysaccharide layer, which appears homogeneous, but with good resolution a filamentous component can be identified. It has been suggested that tropocollagen molecules are the main filamentous component. This lamina has been termed basement membrane, external lamina, boundary layer, or glycocalyx mantle by various authors. The layer is believed to be a product of the adjacent cell. It has an antigenic component which differs from those of the connective tissue. The laminae are permeable to most molecules, but have been reported to act as barriers to the passage of larger particles.

3. In the renal corpuscle a single distinct basal lamina (bl) intervenes between the fenestrated endothelium of the capillary (cap) and the foot processes of the visceral epithelial cells which line the capsular space (cs). Fluid is filtered at a rapid rate across this basal lamina. 3350 ×

4. In epithelial cells the customary appearance of a basal lamina is a thin band which remains at a short distance from the base of the cell. The lamina does not follow the individual folds in the base of the cell, as seen in this epithelium from an arched collecting tubule of the kidney. 7000 ×

5. Renal corpuscle. Filamentous projections from the surfaces of both the footplate of the epithelial cell and from the endothelial cell appear to extend into the basal lamina (arrows). Diaphragms extend across the pores of the fenestrated endothelium and across the gap between foot processes of the epithelial cell. A series of densities creates a line which is within or superimposed upon the cytoplasm between the endothelial pores. 80,000 ×

REFERENCES

GENERAL

Andrew, W. (1966). "Microfabric of Man." Year Book Publ., Chicago, Illinois.

Berkaloff, A. B., Bourguet, J., Favard, P., and Guinnebault, M. (1967). "Biologie et physiologie cellulaire." Hermann, Paris.

Bloom, W., and Fawcett, D. W. (1968). "A Textbook of Histology," 9th ed. Saunders, Philadelphia, Pennsylvania.

Brachet, J., and Mirsky, A. E., eds. (1959–1964). "The Cell," 6 vols. Academic Press, New York.

Cowdry, E. V., ed. (1932). "Special Cytology," 22nd ed., 3 vols. Harper (Hoeber), New York.

De Robertis, E. D. P., Nowinski, W. W., and Saez, F. A. (1965). "Cell Biology" (4th ed. of "General Cytology") pp. 1–446. Saunders, Philadelphia, Pennsylvania.

DuPraw, E. J. (1968). "Cell and Molecular Biology." Academic Press, New York.

Durand, M., and Favard, P. (1967). "La cellule." Hermann, Paris.

Fawcett, D. W. (1966). "An Atlas of Fine Structure. The Cell. Its Organelles and Inclusions." Saunders, Philadelphia, Pennsylvania.

Rhodin, J. A. G. (1961). Cell ultrastructure in mammals. In "The Encyclopedia of Microscopy" (G. L. Clark, ed.), pp. 91–116, Reinhold, New York.

Sandborn, E. B. (1970). Cells and Tissues by Light and Electron Microscopy, Volumes I and II, Academic Press, Inc., New York and London.

Sjöstrand, F. S, (1956). The ultrastructure of cells as revealed by the electron microscope. Intern. Rev. Cytol. 5, 455–533.

Yamada, E., Uchizono, K., and Watanabe, Y., eds. (1968). "Fine Structure of Cells and Tissue Electron Microscopic Atlas," 6 vol. Igaku Shoin, Tokyo.

NUCLEUS

Bernhard, W., and Granboulan, N. (1968). Electron microscopy of the nucleolus in vertebrate cells. In "The Nucleus" (A. J. Dalton and F. Haguenau, eds.), pp. 81–151. Academic Press, New York.

Goldstein, L. (1964). Nucleocytoplasmic relationships. In "Cytology and Cell Physiology" (G. Bourne, ed.), 3rd ed., pp. 560–635. Academic Press, New York.

Hay, E. D. (1968). Structure and function of the nucleolus in developing cells. In "The Nucleus" (A. J. Dalton and F. Haguenau, eds.), pp. 2–80, Academic Press, New York.

Marinozzi, V. (1964). Cytochimie ultrastructurale du nucléole—RNA et protéines intranucléolaires. J. Ultrastruct. Res. 10, 433–456.

Moses, M. J. (1964). The nucleus and chromosomes: A cytological perspective. In "Cytology and Cell Physiology" (G. Bourne, ed.), 3rd ed., pp. 424–558. Academic Press, New York.

MEMBRANE

British Medical Bulletin. (1968). "Structure and Function of Membranes," Vol. 24, No. 2 (with 17 contributors). Med. Dept., Brit. Council, London.

DiCarlo, V. (1967). Ultrastructure of the membrane of synaptic vesicles. Nature 213, 833–835.

DuPraw, E. J. (1968). Structure of the plasma membrane. In "Cell and Molecular Biology," pp. 237–246. Academic Press, New York.

Frisch, D. (1969). A photographic reinforcement analysis of neurotubules and cytoplasmic membranes. J. Ultrastruct. Res., 29, 357–372.

Gasser, H. S. (1955). Properties of dorsal root unmedullated fibers on the two sides of the ganglion. J. Gen. Physiol. 38, 709–728.

Geren, B. B. (1954). The formation from the Schwann cell surface of myelin in the peripheral nerves of chick embryos. Exptl. Cell Res. 7, 558–562.

Korn, E. D. (1966). Structure of biological membranes. Science 153, 1491–1498.

Korn, E. D. (1967). A chromatographic and spectrophotometric study of the products of the reaction of osmium tetroxide with unsaturated lipids. J. Cell Biol. 34, 627–638.

Korn, E. D. (1969). Current concepts of membrane structure and function. Federation Proc. 28, 6–11.

Ling, G. N. (1969). A New Model for the Living Cell: A Summary of the Theory and Recent Experimental Evidence in Its Support. Int. Rev. Cytol. 26:1–61.

Locke, M. ed. (1964). "Cellular Membranes in Development." Academic Press, New York.

Ohnishi, T., and Ohnishi, T. (1962). Extraction of actin and myosin like proteins from erythrocyte membranes. J. Biochem. (Tokyo) 52, 307–308.

Robertson, J. D. (1966b). Current problems of unit membrane structure and contact relationships. In "Nerve as a Tissue" (K. Rodahl and B. Issekutz, eds.), pp. 11–48. Harper (Hoeber), New York.

Sandborn, E. B. (1966). Electron microscopy of the neuron membrane systems and filaments. Can. J. Physiol. Pharmacol. 44, 329.

Sandborn, E. B., Szeberenyi, A., Messier, P. E., and Bois, P. (1965). A new membrane model derived from a study of filaments, microtubules and membranes. Rev. Can. Biol. 24, 243.

Sjöstrand, F. S. (1963). A new ultrastructural element of the membranes in mitochondria and of some cytoplasmic membranes. J. Ultrastruct. Res. 9, 340–361.

Sjöstrand, F. S. (1968). Ultrastructure and function of cellular membranes. In "The Membranes" (A. J. Dalton and F. Haguenau, eds.), pp. 151–210. Academic Press, New York.

Tobias, J. M., Agin, D. P., and Pawlowski, R. (1962). Phospholipid-cholesterol membrane model: Control of resistance by ions or current flow. J. Gen. Physiol. 45, 989–1002.

GOLGI COMPLEX

Beams, H. W., and Kessel, R. G. (1968). The Golgi apparatus: Structure and function. Intern. Rev. Cytol. 23, 209–276.

Caro, L. O., and Palade, G. E. (1964). Protein synthesis, storage and discharge in the pancreatic exocrine cell: An autoradiographic study. J. Cell Biol. 20, 473–495.

Dalton, A. J. (1961). Golgi apparatus and secretions granules. In "The Cell," (J. Brachet, and A. E. Mirsky, eds.), Vol. 2, pp. 603–619. Academic Press, New York.

Peterson, M., and Leblond, C. P. (1964). Synthesis of complex carbohydrates in the Golgi region as shown by radiography after injection of labeled glucose. J. Cell Biol. 21, 143–148.

ENDOPLASMIC RETICULUM AND RIBOSOMES

DuPraw, E. J. (1968). The endoplasmic reticulum. "Cell and Molecular Biology," pp. 446–451. Academic Press, New York.

Siekevitz, P., and Palade, G. E. (1960). A cytochemical study on the pancreas of the guinea pig. V. *In vivo* incorporation of leucine-1-C^{14} into the chymotrypsinogen of various cell fractions. *J. Biophys. Biochem. Cytol.* **7**, 619–630.

GLYCOGEN

Revel, J. P. (1964). Electron microscopy of glycogen. *J. Histochem. Cytochem.* **12**, 104–114.

Revel, J. P. Napolitano, L., and Fawcett, D. W. (1960). Identification of glycogen in electron micrographs of thin tissue sections. *J. Biophys. Biochem. Cytol.* **8**, 575–589.

MITOCHONDRIA

Daems, W. T., and Wise, E. (1966). Shape and attachment of the cristae mitochondria in mouse hepatic cell mitochondria. *J. Ultrastruct. Res.* **16**, 123–140.

Lehninger, A. L. (1965). "The Mitochondrion Molecular Basis. of Structure and Function." Benjamin, New York.

Ohnishi, T., and Ohnishi, T. (1962). Extraction of contractile protein from liver mitochondria. *J. Biochem* (*Tokyo*) **51**, 380–381.

Palade, G. (1953). An electron microscopic study of mitochondrial structure. *J. Histochem. Cytochem.* **1**, 188–211.

MICROTUBULE

Behnke, O. (1964). A preliminary report of "microtubules" in undifferentiated and differentiated vertebrate cells. *J. Ultrastruct. Res.* **11**, 139–146.

Behnke, O., and Forer, A. (1967). Evidence for four classes of microtubules in individual cells. *J. Cell Sci.* **2**, 169–192.

Behnke, O., and Zelander, T. (1967). Filamentous substructure of microtubules of the marginal bundle of mammalian blood platelets. *J. Ultrastruct. Res.* **19**, 147–165.

Burgos, M. H., and Fawcett, D. W. (1955). Studies on the fine structure of the mammalian testis. I. Differentiation of the spermatids in the cat. *J. Biophys. Biochem. Cytol.* **1**, 287–300.

Burton, P. R. (1966). A comparative electron microscopy study of cytoplasmic microtubules and axial unit tubules in a spermatozoon and a protozoan. *J. Morphol.* **20**, 397–424.

deThé, G. (1964). Cytoplasmic microtubules in different animal cells. *J. Cell Biol.* **23**, 265–275.

Freed, J. (1965). Microtubules and saltatory movements of cytoplasmic elements in cultured cells. *J. Cell Biol.* **27**, 29-A (abstr.).

Gall, J. G. (1966). Microtubule fine structure. *J. Cell Biol.* **31**, 639–643.

Sandborn, E. B., Koen, P. F., McNabb, J. D., and Moore, G. (1964). Cytoplasmic microtubules in mammalian cells. *J. Ultrastruct. Res.* **11**, 123–138.

Slautterback, D. B. (1963). Cytoplasmic microtubules. I. Hydra. *J. Cell Biol.* **18**, 367–388.

Wolfe, S. L. (1965). Isolated microtubules. *J. Cell Biol.* **25**, 408–413.

LYSOSOMES, LIPOFUCHSIN GRANULES, AND MICROBODIES (PEROXYSOMES)

Afzelius, B. A. (1965). The occurrence and structure of microbodies. A comparative study, *J. Cell Biol.* **26**, 835–843.

Bowers, W. E., and deDuve, C. (1967a). Lysosomes in lymphoid tissue. II. Intracellular distribution of acid hydrolases. *J. Cell Biol.* **32**, 339–349.

deDuve, C. (1963). The lysosome concept. *Ciba Found. Symp., Lysosomes*, pp. 1–31.

deDuve, C., and Baudhuin, P. (1966). Peroxisomes. *Physiol. Rev.* **46**, 323.

Novikoff, A. B. (1961). Lysosomes and related particles. *In* "The Cell" (J. Brachet and A. E. Mirsky, eds.), Vol. 2, pp. 423–488. Academic Press, New York.

Shnitka, T. K. (1966). Comparative ultrastructure of hepatic microbodies in some mammals and birds in relation to species differences in uricase activity. *J. Ultrastruct. Res.* **16**, 598–625.

FILAMENTS

Cloney, R. A. (1966). Cytoplasmic filaments and cell movements: Epidermal cells during ascidian metamorphosis. *J. Ultrastruct. Res.* **14**, 300–328.

McNabb, J. D., and Sandborn, E. B. (1964). Filaments in the microvillous border of intestinal cells. *J. Cell Biol.* **22**, 701–704.

Sandborn, E. B. (1966). Electron microscopy of the neuron membrane systems and filaments. *Can. J. Physiol. Pharmacol.* **44**, 329–338.

Zwillenberg, L. O. (1965). Filament-carrying tubules demonstrated by negative staining in various mammalian cell types. *Z. Zellforsch. Mikroskop. Anat.* **66**, 415–426.

MICROVILLI

Ito, S. (1965). The enteric surface coat on cat intestinal microvilli. *J. Cell. Biol.* 27, 475–491.

McNabb, J. D., and Sandborn, E. B. (1964). Filaments in the microvillous border of intestinal cells. *J. Cell Biol.* **22**, 701–704.

Overton, J. (1968). Localized Lanthanum staining of the intestinal brush border. *J. Cell Biol.* **38**, 447–452.

Sjöstrand, F. S., and Zetterquist, H. (1957). Functional changes of the free cell surface membrane of the intestinal absorbing cell. *Proc. Reg. Conf.* (*Eur.*) *Electron Microscopy, Stockholm*, 1956 pp. 150–151. Academic Press, New York.

JUNCTIONAL COMPLEX

Bullivant, S., and Lowenstein, W. R. (1968). Structure of coupled and uncoupled cell junction. *J. Cell. Biol.* **37**, 621–632.

Coggeshall, R. E. (1966). A fine structural analysis of the epidermis of the Earthworm Lumbricus terrestris L. *J. Cell Biol.* **28**, 95–108.

Farquhar, M. G., and Palade, G. E. (1963). Junctional complexes in various epithelia, *J. Cell Biol.* **17**, 375–412.

Trelstad, R. L., Revel, J. P., and Hay, E. D. (1966). Tight junctions between cells in the early thick embryo as visualized with the electron microscope. *J. Cell Biol.* **31**, C6–C10.

CENTRIOLES

André, J. (1964). Le centriole et la région centrosomienne. *J. Microscopie*, 3, 23.

Bernhard, W., and deHarven, E. (1960). L'ultrastructure du centriole et d'autres éléments de l'appareil achromatique. *Proc. 4th Intern. Congr. of Electron Microscopy, Berlin 1958.* Vol. II, pp. 217–227. Springer, Berlin.

deHarven, E. (1968). The centriole and the mitotic spindle. *In* "The Nucleus" (A. J. Dalton and F. Haguenau, eds.), pp. 197–227. Academic Press, New York.

Szollosi, D. (1964). Centrioles, centriolar satellites and spindle fibers. *Anat. Record* **148**, 343.

CELL IN MITOSIS

Buck, R. C., and Krishan, A. (1964). Site of membrane growth during cleavage of amphibian epithelial cells. *Exptl. Cell Res.* **38**, 426–428.

Krishan, A., and Buck, R. C. (1964). Structure of the mitotic spindle in L strain fibroblasts. *J. Cell Biol.* **24**, 433–444.

Ledbetter, M. C. (1967). The disposition of microtubules in plant cells during interphase and mitosis. *Symp. Intern. Soc. Cell Biol.* **6**.

Levine, L., ed. (1963). "The Cell in Mitosis" Academic Press, New York.

Robbins, E., and Gonatas, N. K. (1964). The ultrastructure of a mammalian cell during the mitotic cycle. *J. Cell. Biol.* **21**, 429–463.

CILIA AND FLAGELLA

Gibbons, I. R. (1967). The structure and composition of cilia. *Symp. Intern. Soc. Cell Biol.* **6**, 99–114.

Roth, L. E., and Shigenaka, Y. (1964). The structure and formation of cilia and filaments in rumen protozoa. *J. Cell Biol.* **20**, 249–270.

Satir, P. (1965). Studies on cilia. II. Examination of the distal region of the ciliary shaft and the role of the filaments in motility. *J. Cell. Biol.* **26**, 805–834.

Steinman, R. M. (1968). An electron microscopic study of ciliogenesis in developing epidermis and trachea in the embryo of Xenopus larvis. *Am. J. Anat.* **122**, 19–55.

EXTERNAL COATS

Bennett, H. S. (1963). Morphological aspects of extracellular polysaccharides. *J. Histochem. Cytochem.* **11**, 14–23.

Brandt, P. W. (1962). A consideration of the extraneous coats of the plasma membrane. Symposium on the plasma membrane. *Circulation* **26**, 1075–1091.

Ito, S. (1965). The enteric surface coat on cat intestinal microvilli, *J. Cell Biol.* **27**, 475–491.

Pease, D. C. (1966). Polysaccharides associated with the exterior surface of epithelial cells: Kidney, intestine, brain, *J. Ultrastruct. Res.* **15**, 555–588.

Pierce, G. B., Jr., Beals, T. F., Sri Ram, J., and Midgley, A. R. (1964). Basement membranes. IV. Epithelial origin and immunologic cross reactions. *Am. J. Pathol.* **45**, 929–961.

FUNDAMENTAL TISSUES

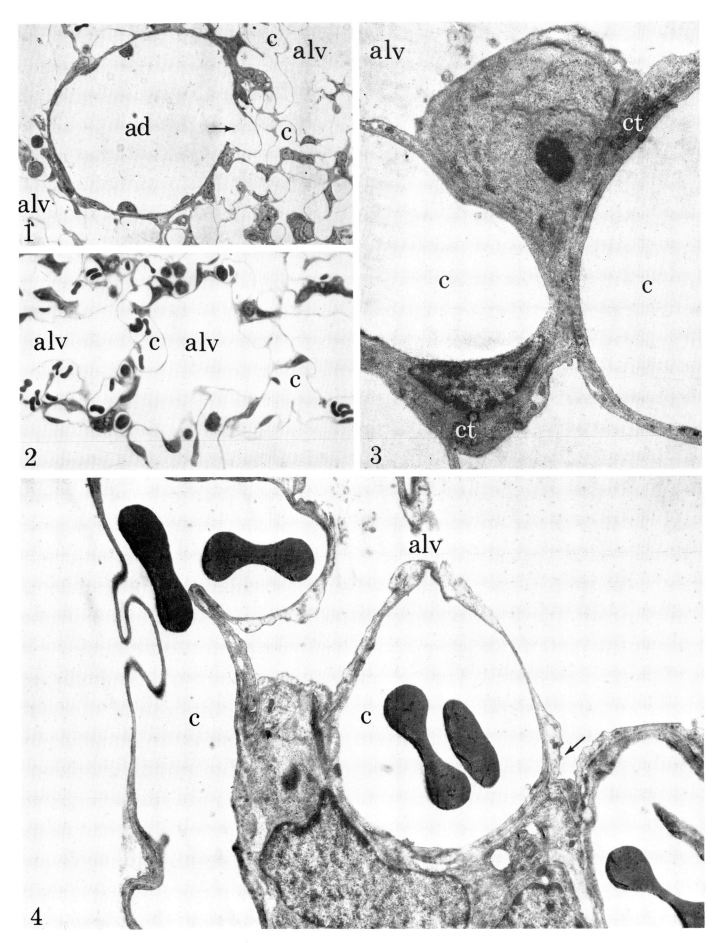

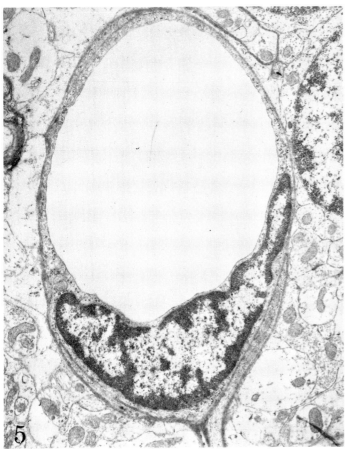

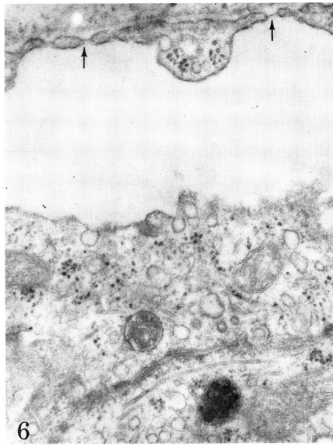

Epithelia

Cells, adapted to their environments, cover all surfaces of the body. All conducting passages in the body are lined by epithelial cells arranged in one or more layers. These have been classified as squamous, cuboidal, or columnar, depending on the height of the cells. The different types will be encountered in various organs.

The simple squamous cell epithelium provides a single surface layer of flattened cells. In addition this thin layer allows rapid transfer of gases and fluids through and between the cells. Blood vessels are lined by squamous cells in a single continuous layer. These cells, of mesodermal origin, constitute endothelium.

The respiratory system has been found ideal for presenting many features of fundamental tissues. The lung, where the gaseous interchanges of oxygen and carbon dioxide take place, provides simple squamous epithelial cells and endothelial cells in thin layers. This barrier between air and circulating blood allows a rapid exchange of gases between the two.

1. L. Lung. The alveolar duct (ad),with a thin layer of connective tissue containing collageneous and elastic fibers supporting its simple squamous epithelial lining, is in the transitional zone wherein conducting passages give way to the tissue specialized for the gaseous exchange. This transition is illustrated in the alveolar outpouching (arrow) of the thin epithelial lining of the duct wall. Surrounding the alveolar duct are a number of alveoli and capillaries. Thin bands of cytoplasm form the barriers between air in the alveolar spaces (alv) and blood in the capillary (c) lumen. The capillaries are empty in this perfused lung. Flattened endothelial cells form the wall of the capillaries. Larger epithelial cells protrude out into the more irregularly shaped alveolar air spaces from clefts between the capillaries. 500 ×

2. L. A great amount of surface contact is provided for respiratory exchange between the capillaries (c), in which some blood cells have remained, and the larger irregular alveolar (alv) air spaces. 500 ×

3. The alveolar (alv) epithelial cell, located in the angle between capillaries (c), has a pale nucleus with a large, dense nucleolus. The endothelial cell nucleus is elongated and conforms to the capillary wall. Small, dense granules are found in both types of cells. At this magnification separate layers of cytoplasm are difficult to identify in the septa between capillary and alveolus. Small collections of densely stained connective tissue (ct) containing collagen fibrils and elastic fibers are located in the angles between the epithelial and the endothelial cells. 8700 ×

4. The cytoplasm of the endothelium of the smaller capillaries (c) can be distinguished from that of the larger alveolar (alv) epithelial cells. Alveolar or septal cells occupy niches between capillaries. Cytoplasmic extensions of these cells completely surround the alveolar space. Small, dense bodies and a few vacuoles are present in these cells. A few short microvilli are present on the surface of the alveolar epithelium (arrow). 6000 ×

5. Capillary—cerebellum. A single flattened endothelial cell is able to completely encircle the lumen of a capillary. In the central nervous system, with the exception of neurosecretory appendages, the endothelial cell cytoplasm is rarely of the very thin, fenestrated type. The elongated nucleus occupies a thickened part of the capillary wall. The cell rests on a well-defined basal lamina. 10,000 ×

6. Thyroid. One wall of a capillary is fenestrated (arrows), while the other appears quite thick. The thicker portion represents opposite borders of a single endothelial cell which are united by desmosomes. A number of vesicles, microtubules, ribosomes, and membrane-bound granules of varying density are found in the cytoplasm. 54,000 ×

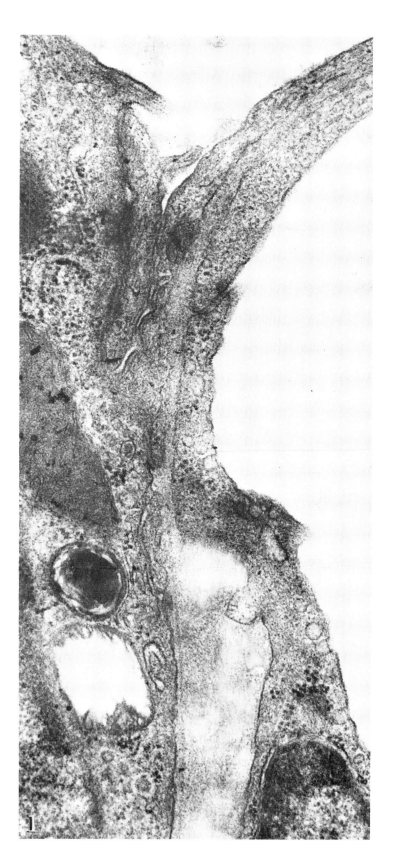

Epithelia—Lung

1. The thin septum between the alveolus and the capillary is made up of cytoplasmic layers of both the epithelial and the endothelial cell. Thin basal laminae are interposed between the two cells. Mitochondria in the epithelial cell stain intensively. Granules with varying density and internal structure are occasionally seen in alveolar cells. Vacuoles contain threads of moderately dense material. The alveolar epithelial cells have a few short microvilli and operculae on their surfaces. These cells *in situ* are apparently not phagocytic. 75,000 ×

Epithelium

2. L. The trachea is a hollow organ lined by columnar epithelial cells on a wide band of loose connective tissue. This, in turn, is supported at its periphery by a thin layer of dense connective tissue which blends into a wide band of cartilage. The cartilage cells appear as lighter spaces in a moderately dense homogeneous matrix. The narrow interrupted line of extremely dense granules in the loose connective tissue is a thin layer of elastic fibrils (arrows) sectioned transversely. 110 ×

3. L. Ciliated and nonciliated cells are included in the epithelial lining of the trachea. Although each individual cell reaches the basal lamina, the differing levels of their nuclei have created the false impression that there is more than one layer of cells. This type has been termed pseudostratified columnar epithelium. The loose connective tissue contains capillaries and a variety of cells including fibroblasts, small lymphocytes, plasma cells, macrophages, and others. Some of the lymphocytes and more densely stained polymorphonuclear leukocytes have invaded the epithelial layer. 450 ×

4. Adjacent cells of the tracheal epithelium have a remarkable variation in their apices. Some are nonciliated, but bear short, thick apical microvilli, while among the cilia on the apices of the other cells there are many of the long, thin microvilli. Included in any one section is only a small fraction of the very large number of these long, thin microvilli and cilia on a single cell. Basal bodies of the cilia are seen within the apex of the cell. The apices of the nonciliated cells often protrude into the lumen. Both types of cells contain some dense granules. The nonciliated cells are believed to be secretory in nature. 8100 ×

5. The apical protrusion of these nonciliated cells appear much taller than they actually are since the cells have been sectioned obliquely. The tips of the microvilli have a granulated appearance (arrows), and this cell contains a number of large, dense apical granules. A number of cilia in transverse or oblique sections within the lumen of the trachea and basal bodies in cell apices can be seen. A dense granule is located in the apex of one of the ciliated cells. 25,000 ×

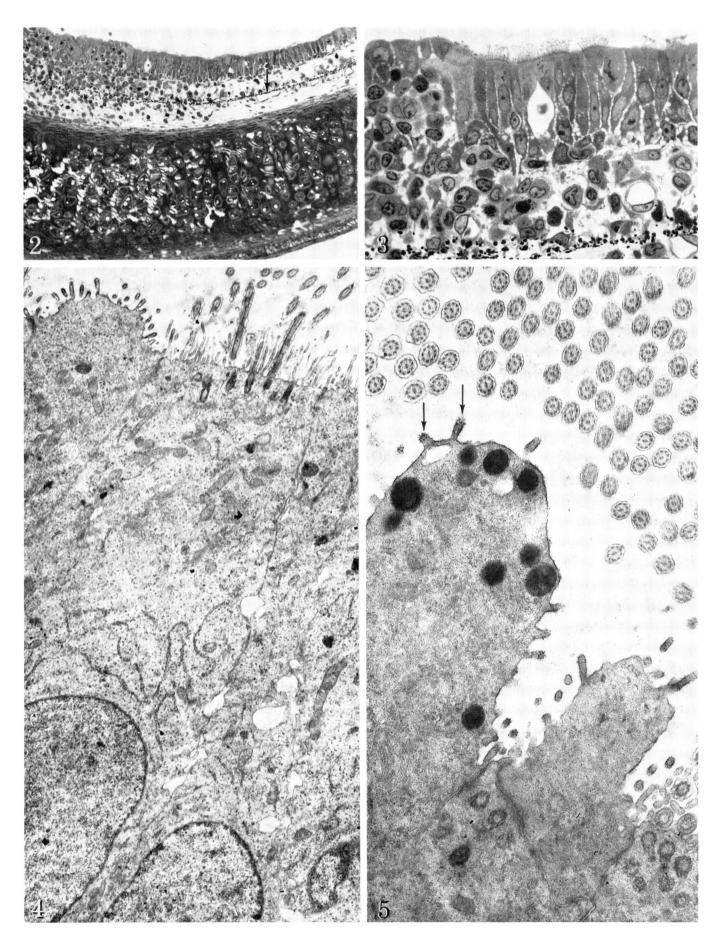

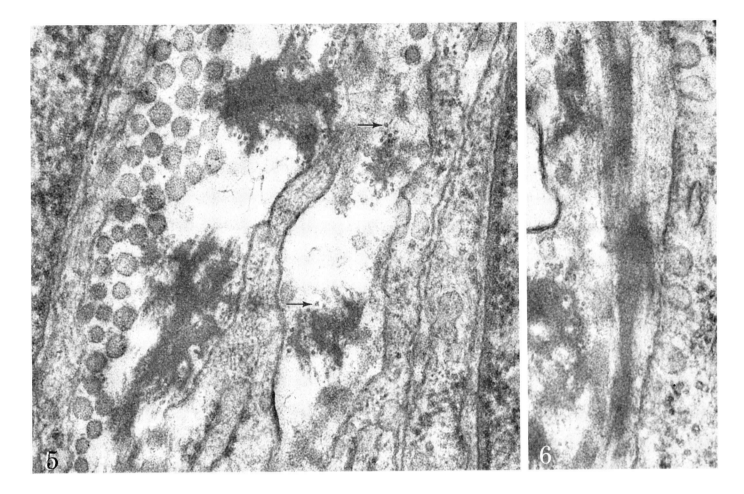

Connective Tissue—Fibroblasts, Collagen, and Elastic Fibers

1. L. The vas (ductus) deferens. In some locations extremely cellular connective tissue is found. The lamina propria of the wall of the vas deferens is made up of a thick layer of fibroblasts, collagen, and densely stained elastic fibers. 600 ×

2. Collagen fibrils, located between the elongated fibroblasts, are sectioned transversely and longitudinally. There is a narrow dense band within the periphery of the fibroblast nuclei. A large Golgi complex (G) and a number of surface membrane invaginations are encountered in the fibroblasts. Elastic fibers, reportedly difficult to demonstrate, are very prominently displayed by the methods employed for the preparation of the material for this study. These very densely stained elastic fibers with irregular outlines are scattered between the collagen fibrils and the fibroblasts. Dense granules, of an intensity equaling that of the elastic fibers, are located in the cytoplasm of several of the fibroblasts. Although little is known of the origin of elastic fibers, it is commonly believed that elastin is synthesized by fibroblasts. The dense granules within the fibroblasts are quite possibly elastin or a precursor about to be secreted into the extracellular space. 16,000 ×

3. An enlargement of an intracytoplasmic granule, elastic fibers, and collagen from Fig. 2 demonstrates one of the intracellular dense bodies and the extracellular elastic fibers. 32,000 ×

4. In an enlargement of another part of Fig. 2 the intracellular dense body is surrounded by a filamentous network, a part of which appears to contact the plasma membrane (arrow). 35,500 ×

5. When adequately preserved, elastic fibers appear as irregular dense, intercellular masses in the electron microscope. The elastin appears as a homogeneous dense material in variable quantity with a markedly irregular outline. Each dense mass is bordered by closely applied filaments with a diameter of approximately 60 Å. Many of these filaments have a distinct central lumen (arrows). The transversely sectioned collagen fibrils, which, with the elastic fibers, are located between the fibroblasts, range in size from 600 to 1200 Å in diameter. The fibroblasts display a variable quantity of moderately dense material internal to the inner nuclear membrane. The density is diminished at the nuclear pore. 93,000 ×

6. Internal elastic lamina—arteriole. A longitudinal view of an elastic fiber demonstrating the accompanying filaments in parallel 88,000 ×

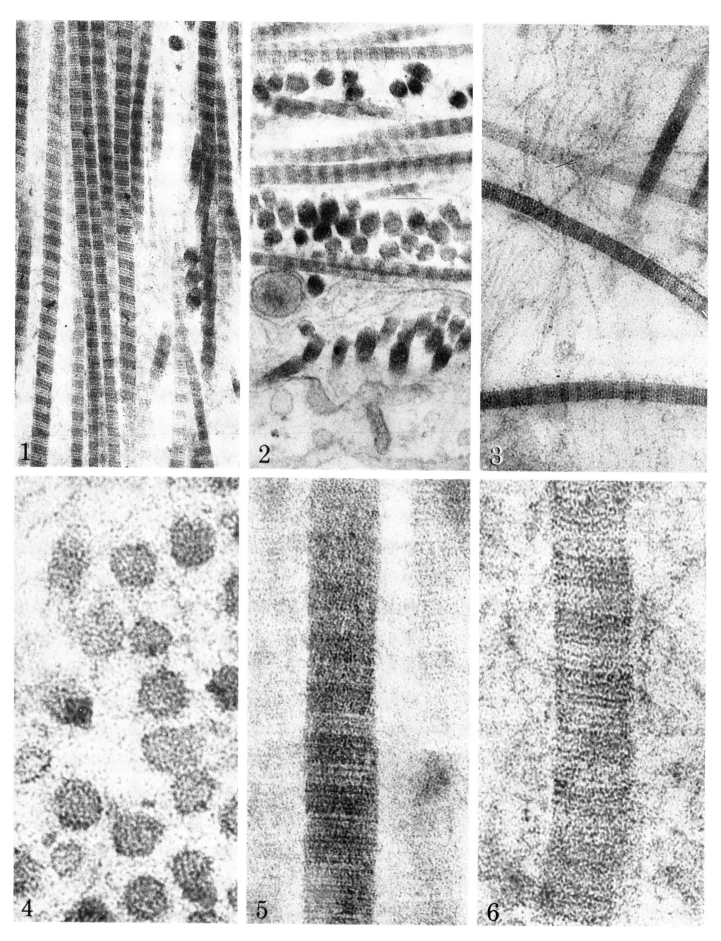

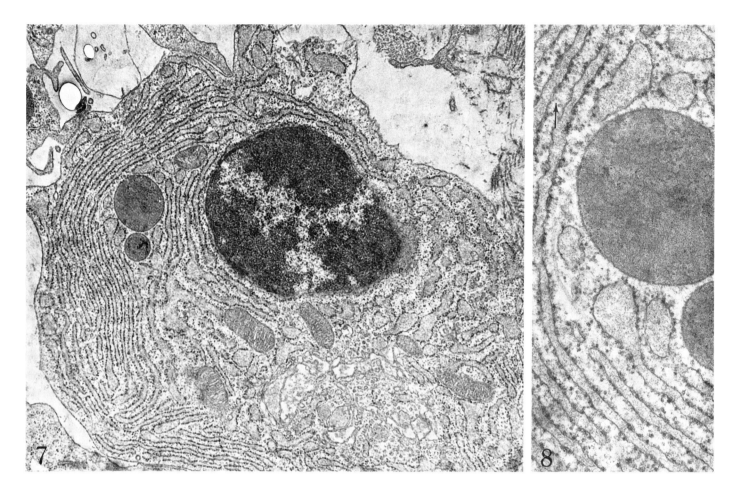

Connective Tissue—Collagen Fibrils and Reticular Filaments

Collagen fibrils give the tissues their tensile strength. Consequently, there is a great variation in quantity, depending upon the stresses involved.

1. The collagen fibrils are circular in transverse section. In longitudinal section they display a distinct banding. The light and the dark bands are again subdivided by dense transverse lines. This banding is believed to be produced by a quarter staggering of the tropocollagen molecules, each of which is approximately 2800 Å in length. The banding does not correspond exactly with quarters of the tropocollagen molecule, probably because of variable overlapping and a helical disposition of the molecules. 85,000 ×

2. Collagen fibrils, as seen in longitudinal and transverse section, vary in diameter and in intensity of heavy metal staining. The latter depends upon the band, which might be included in the section. The large membrane-bound granule and the vesicles are intracellular. 85,000 ×

3. In loose connective tissue from the choriovascular layer of the retina, extracellular reticular filaments form a multidirectional network among the scattered collagen fibrils. Some of these filaments appear striated. 85,000 ×

4. In transverse sections, at high magnification, the collagen fibrils can be seen to be made up of circular subunits. 310,000 ×

5. At high magnification, longitudinally oriented subunits of the collagen can be detected in the fibril. 275,000 ×

6. The dense bands in the collagen fibrils, when adequately resolved, are made up of longitudinally oriented densities placed side by side (arrow). 310,000 ×

Loose Connective Tissue, Plasma Cell—Trachea

The plasma cell, the origin of which is still debated, is highly specialized for the synthesis of antibodies. This cell is rarely found except in lymphoid tissue, the lining of the respiratory and intestinal tracts, and in serous membranes.

7. A view of the eccentric nucleus and a large Golgi apparatus with lightly stained content is obtained in this section. This illustrates the cisternae of the granular endoplasmic reticulum in long profiles. A few large, homogeneous, dense granules are found in some plasma cells of the rat. It is generally accepted that immune globulins are synthesized and stored in the granular reticulum of plasma cells. 11,000 ×

8. At a higher magnification the large dense granules are bounded by a smooth-surfaced membrane. Cytoplasmic microtubules are located among the cisternae. Occasionally, filaments project out from and sometimes appear to interconnect the membranes of adjacent organelles (arrow). Whether the content of the dense granules plays a role in the process of synthesis or represents another product for export is not known. 42,500 ×

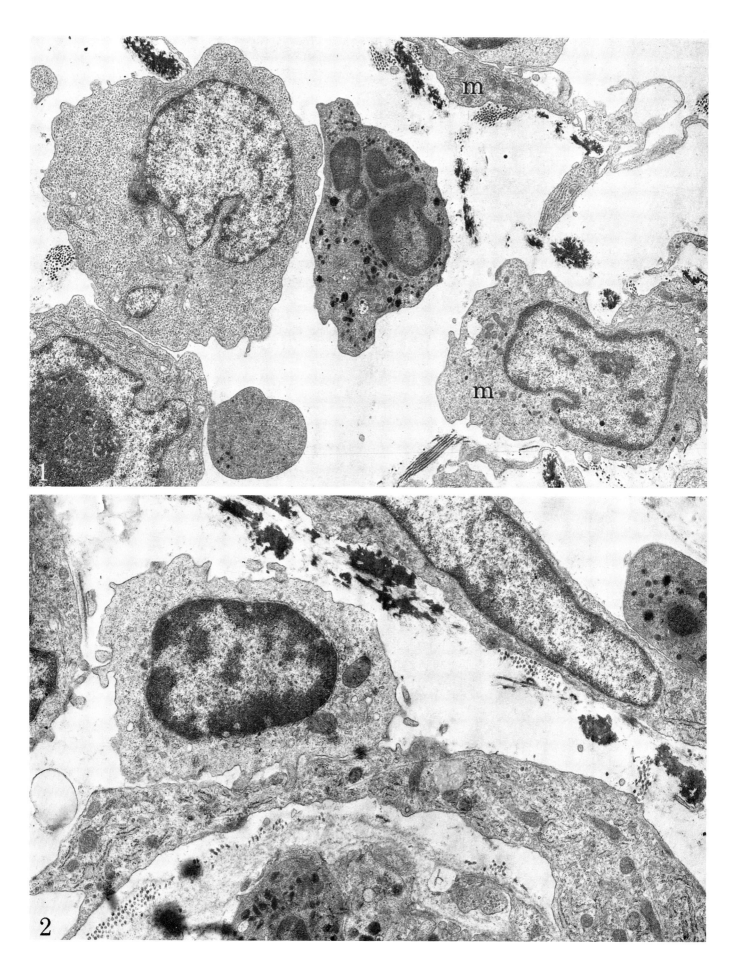

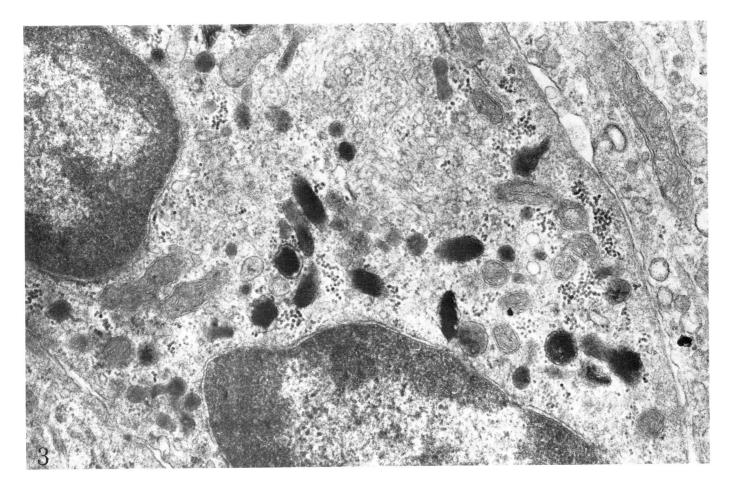

Loose Connective Tissue, Polymorphonuclear Heterophil—Trachea

1. The large circular cell in which the two poles of a kidney-shaped nucleus have been transsected unequally is a monocyte. Vacuoles and small, dense granules are seen in the large Golgi zone. The nucleus is moderately pale with a thin rim of peripheral chromatin. This peripheral chromatin is more distinct than the few small central masses. These cells, apparently capable of survival for long periods of time, are readily transformed into macrophages when adequately stimulated. The cell with the large nucleolus contains a considerably increased quantity of granular endoplasmic reticulum. This could indicate that a differentiation into another cell type is taking place. Another cell (m) with a number of small granules and processes is typical of the macrophage. The cell process with a number of long pseudopods is also a part of a macrophage. A polymorphonuclear heterophil is sectioned through its nucleus and another, through the cytoplasm alone. The density of the background cytoplasm is a distinctive feature of the polymorph. 6000 ×

2. The small cell in the center of the field has a dense nucleus, a moderate amount of granular reticulum, a medium-sized Golgi, and numerous pseudopodial cytoplasmic projections. The appearance of the nucleus suggests that this is a wandering lymphocyte. There is an unusual amount of endoplasmic reticulum in the cytoplasm. Some of the cytoplasmic projections fit into deep indentations in the fibroblast. The fibroblast contains a pale, circular lipid droplet. Another elongated fibroblast is located near the band of irregular dense elastic fibers. 12,500 ×

3. The polymorphonuclear heterophil nuclei contain a wide peripheral band of dense chromatin with the granules often arranged in parallel lines. The cytoplasm contains a great deal of smooth endoplasmic reticulum, numerous glycogen granules and many larger, irregular, membrane-bound granules of variable density. The glycogen granules are rounded and do not appear in rosettes, as seen in the liver parenchymal cells.

The membrane-bound granules of the polymorphonuclear heterophils display a variety of internal structural appearances. These range from multiple indistinct tubular units to "fingerprint" arrangements in a finely granular matrix. These granules have been shown, in the rabbit, to contain a number of hydrolytic enzymes, which probably account for their ability to phagocytose and to destroy bacterial invaders. These enzymes include the antimicrobial agent phagocytin as well as acid phosphatase, nucleotidase, ribonuclease, deoxyribonuclease, glucuronidase, cathepsin, cationic proteins, and acid mucosubstances. Recent evidence suggests that the enzymes in the granules may vary at different stages of development of the heterophil and that only in early stages of development are the hydrolytic enzymes present. The less dense granules with a granular or tubular type of substructure in mature cells have been found to contain alkaline phosphatase and lysozyme but no acid phosphatase. A third type of granule which contains acid phosphatase cationic proteins and acid mucosubstances has been suggested from studies of cell fractions. The heterogeneous granules seen in this figure, with homogeneous dense material superimposed on the less dense granular substructure, possibly represent this third type of granule. 11,500 ×

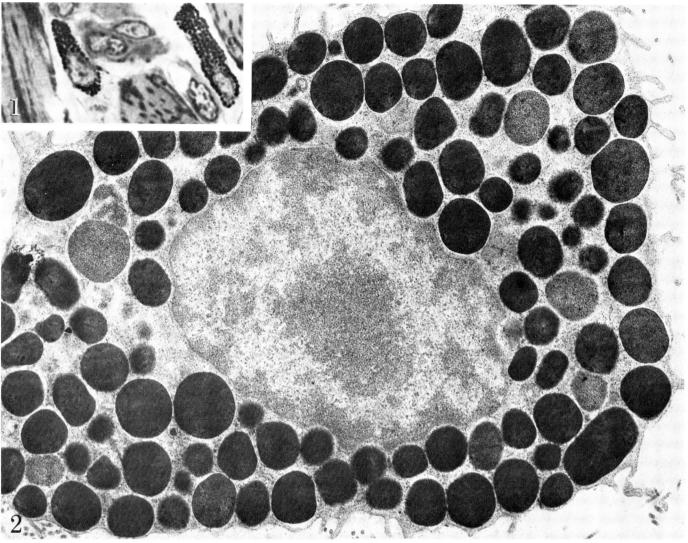

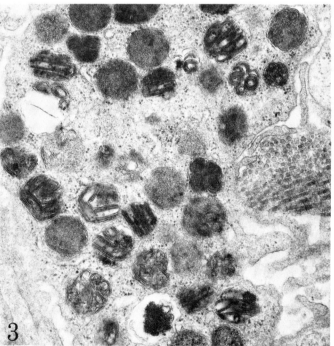

Connective Tissue—The Mast Cell

The mast cell is distinguished in toluidine blue sections by a slight metachromasia of its numerous, moderately large, dense granules. In the rat these cells are not uncommon in most organs, but are more frequently encountered in certain locations such as muscle, the cranial aponeurosis, the thyroid, the mammary gland, and in the intestine. The shape of the cell reflects the compartment in which it is located. In muscle they are often elongated with their moderate-sized, pale nuclei at one pole. The mast cell is responsible for the synthesis of the anticoagulant, heparin, and the powerful vasodilator, histamine. When mast cells are partially degranulated, the metachromasia of the granules becomes more evident in sections prepared in this fashion.

1. L. Tongue. Elongated mast cells with numerous discrete, dense cytoplasmic granules are found between muscle fibers. 1000 ×

2. Thyroid. Although the chromatin is present in considerable concentration, the nucleus appears pale. This is explained by the very fine granularity of the chromatin. The cytoplasmic surface projections resemble the microvilli of epithelial cells. The granules are either very dense or are moderately stained with a fibrillar background. The latter, we have found, is the general appearance of the granules in "degranulated" cells of light microscopy. 16,000 ×

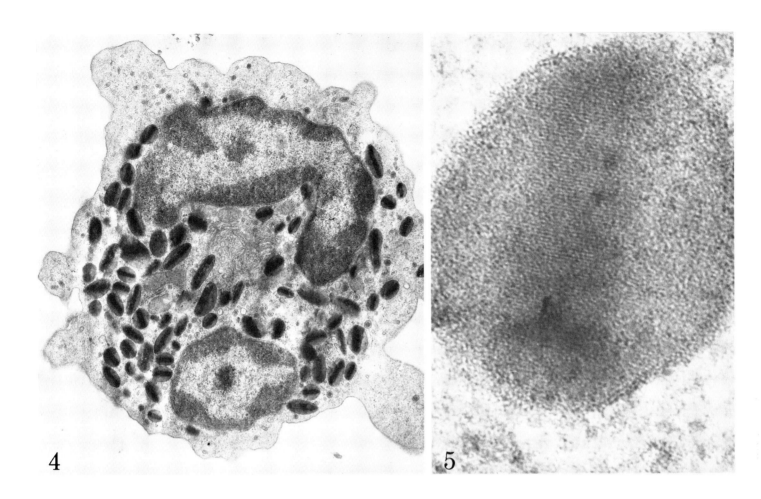

4

5

Mast Cell—Human Liver

3. Mast cell granules in the human are quite unlike those of the rat. The granules vary in density and in internal substructure. The individual granules, each surrounded by a membrane, vary from dense,with only a faintly discernible background to ever-increasing complexities of lines and whorls. These have been termed fingerprint types of granules. The cell border is irregular, and the cytoplasmic projections are large. 27,500 ×

Connective Tissue—The Eosinophil

Eosinophils leave the capillaries to migrate through the tissues. They have long been known to increase in number in allergic reactions. Recent evidence indicates that they phagocytose and break down antigen-antibody complexes. They contain peroxidase, ribonuclease, aryl sulfatase, cathepsin, β-glucuronidase, and alkaline phosphatase.

4. Trachea. In loose connective tissue the cell is rounded with bulbous surface projections. The nucleus has a wide peripheral band of finely granular chromatin, and a large central Golgi is present in the cytoplasm. A central bar of very dense material is clearly outlined in each of the many dense granules in the cytoplasm. The peripheral cytoplasm contains a number of smaller granules of lesser density. 12,500 ×

5. Parallel lines are quite evident in this section, which has passed obliquely through the central band of one of the eosinophil granules. The parallel lines have been interpreted as evidence of the presence of a crystalline material in the granule. In the adjacent granule, where the band has been sectioned transversely, there is an appearance of separate, linear substructural units at right angles to each other, thus creating a woven appearance (arrow). 230,000 ×

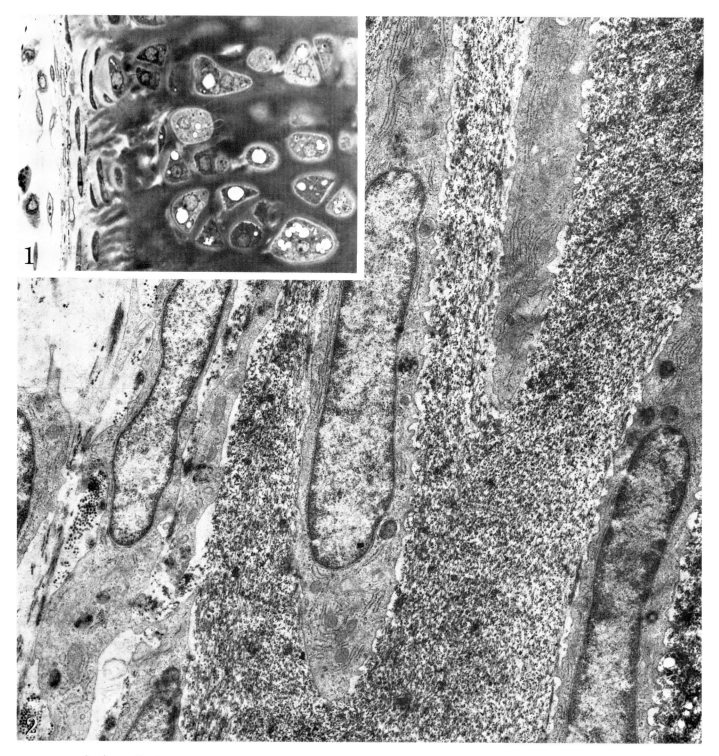

Cartilage—Trachea

1. L. The perichondrium, a thin layer of fibroblasts with collagen and a few densely stained elastic fibers, gives way to chondroblasts around which a glycoprotein ground substance is laid down. This chondromucoid contains chondroitin sulfate A and C in a 4:1 ratio. As the cartilage matures, the cells, now termed chondrocytes, become rounded or angular. The chondrocytes contain small granules and clear spaces. A lipid content has been removed from the vacuoles during the tissue preparation. 500 ×

2. In a field which includes the perichondrium, bundles of collagen and homogeneous, dense elastic fibers surrounding the fibroblasts become impregnated with deposits of chondromucoid A few very dense granules are located in the fibroblasts and the chondroblasts. Elastin can be rcognized as the scattered densities in the ground substance. The chondroblasts contain considerable quantities of granular reticulum and a number of dense bodies. The cell border has developed an irregular outline with short cytoplasmic projections extending out into the ground substance. 9700 ×

The cartilage of the trachea is of the hyaline type, but there is evidently an element of elastic fibers in it. Increased amounts of the elastic fibers typify elastic cartilage, while fewer chondrocytes and more collagen are found in fibrocartilage.

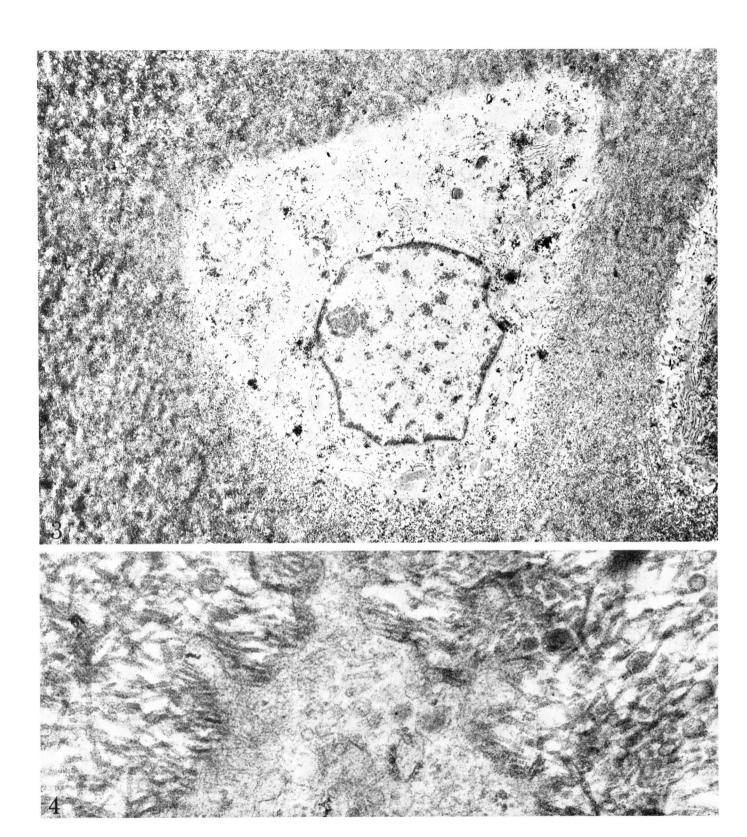

Cartilage—Trachea

3. In the mature chondrocyte the nucleus has an angular form with discrete accumulations of chromatin scattered throughout. Accumulations of the dense, irregular, homogeneous material are scattered through the cytoplasm. There appears to be no sharply defined cell border. 6600 ×

4. In obliquely sectioned borders of the mature chondrocyte there is no evidence of a plasma membrane. There is a dense network of filaments which apparently blends into the thicker, denser fibers which support the extracellular cartilage matrix. 78,000 ×

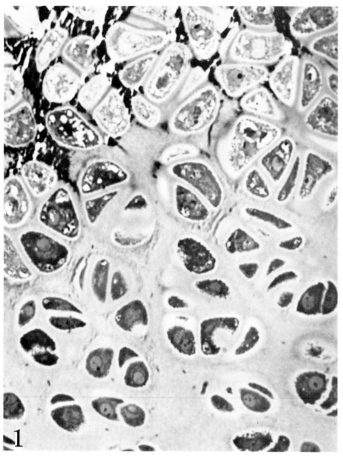

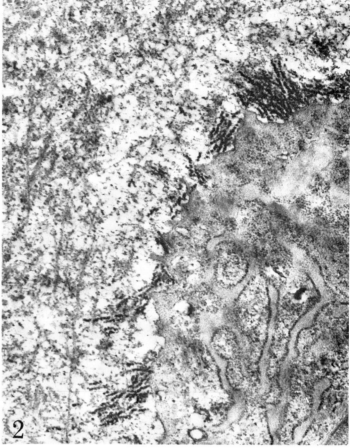

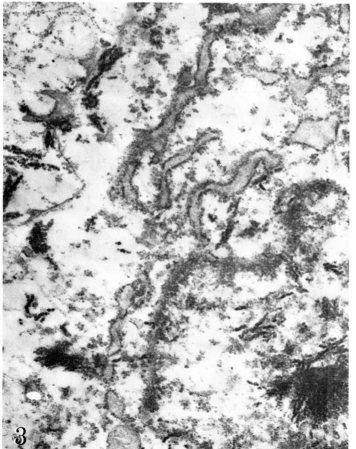

Calcification of Cartilage—Os Priapi

Cartilage participates in the elongation of bones by a rapid proliferation in specific end zones. At a distance from the zone of proliferation a calcification of the mature cartilage takes place. This provisional calcification is followed by a complicated removal and a reconstruction process wherein a more organized type of bone replaces the calcified cartilage.

1. L. Os priapi in a young rat. Chondrocytes in the hyaline cartilage have a basophilic cytoplasm. As the cells age, they pass through phases of maturation, hypertrophy, and degeneration. As the chondrocytes degenerate, the cytoplasm becomes less basophilic and progressively more vacuolated. The pale, homogeneous cartilaginous matrix around the degenerating cells becomes infiltrated with densely stained calcium phosphate and calcium carbonate deposits. 450 ×

2. As the cell degenerates, the cisterns of the granular endoplasmic reticulum becomes more dilated with faintly defined free ribosomes occupying the space between them. A considerable quantity of the coarser chains of calcium crystals is seen along the cell surface. 28,000 ×

3. As the intercellular laminae of the calcified cartilage is dissolved in the development of the lacunae, and as degeneration of the cell progresses, the background chondrocyte cytoplasm becomes clearer. The free ribosomes gradually disappear, and calcium deposits appear among the scattered cisternae of the endoplasmic reticulum. Crystals of calcium appear in the nucleus as well. 28,000 ×

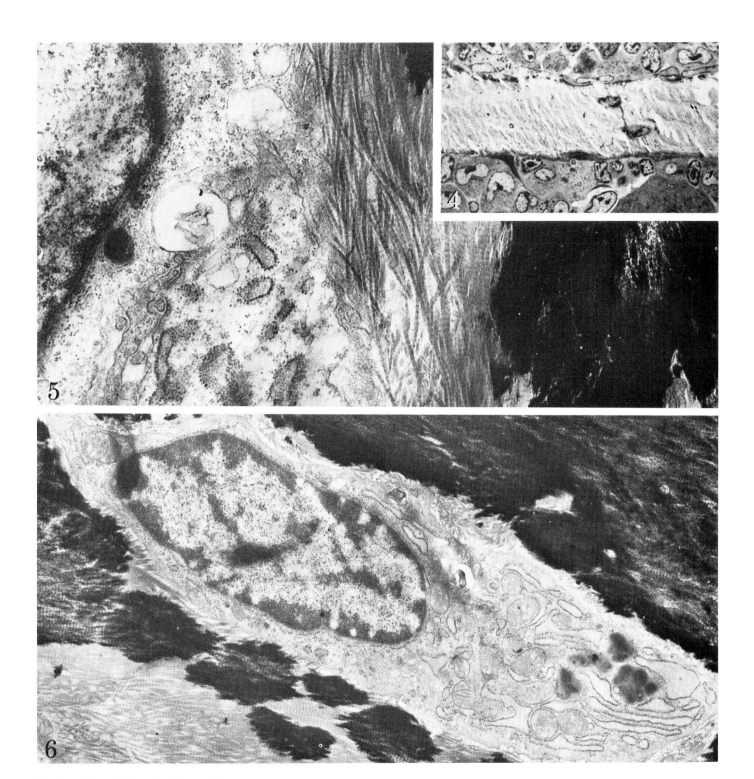

The Osteoblast—Trabecula of Bone—Sternum

In the constant process of bone rebuilding the osteoblasts are responsible for the development of the collagenous stroma and for the extracellular deposition of calcium.

4. L. The cells with the single elongated nuclei and varying amount of cytoplasm along one border of the bone are osteoblasts. Some of the osteoblasts contain a number of small, dense granules. As the bone develops, the osteoblasts become trapped and are then termed osteocytes. The lacunae in which two osteocytes are located are interconnected by a canaliculus. A large multinucleated cell in contact with the other border of the bone is an osteoclast. These cells remove bone in the continuous destruction and rebuilding process which goes on throughout the bone during life. 720×

5. The osteoblast has an abundant supply of granular endoplasmic reticulum with moderately dilated cisternae. A dense granule is located near the nucleus. Collagen fibrils in a ground substance are laid down in the extracellular space by the osteoblast. Calcium is deposited on the fibrous stroma a short distance from the cell border. 25,600×

6. As the calcification of the stroma continues, the osteoblast becomes surrounded by bone, wherein it becomes an osteocyte. The density of the bone is greatest at the border of the developing lacuna. Dense, homogeneous granules are seen in the osteocyte. 13,500×

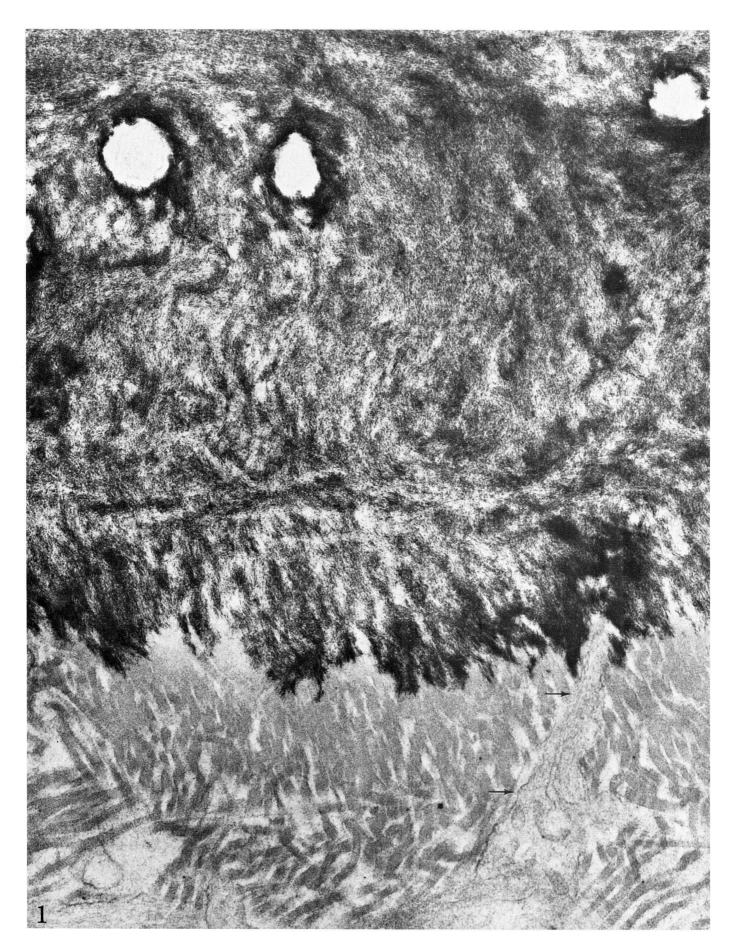

1

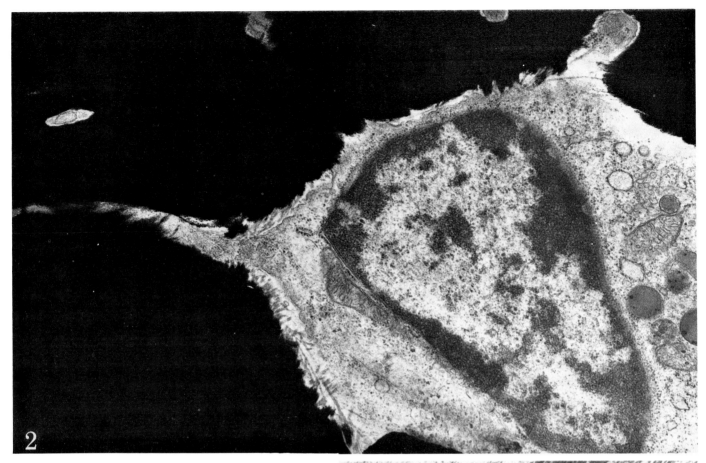

Bone

1. The calcium is deposited as fine, elongated crystals on the fibrous stroma. Cytoplasmic processes from neighboring osteoblasts extend between collagen fibrils into the developing bone (arrows). In transverse section the cytoplasmic processes can be seen within canaliculae in the bone. The cytoplasmic processes are surrounded by a membrane and contain numerous filaments, as demonstrated both in longitudinal and in transverse sections. There appears to be a homogeneous density superimposed upon the calcium crystals along the border of the bone. This increased density is evident in the bone surrounding the canaliculi. 53,000×

Bone—Osteocyte

2. The osteocyte is completely encompassed by bone. Long cytoplasmic processes occupy the canaliculi. A narrow uncalcified zone surrounds the cell and its processes. It has been shown that these form an anastomosing series of channels between lacunae in the bone. Deposition of calcium no longer occurs. The cell retains a moderate amount of endoplasmic reticulum, mitochondria, and a number of homogeneous, dense, membrane-bound granules. 25,500×

3. The bone mineral consists of a mass of hydroxyapatite crystals deposited within and around the collagen fibrils. The crystals range from 15 Å to 75 Å in thickness and are up to 1500 Å in length. Some of the crystals are curved. In most instances the direction of the crystals in the collagen parallel the fibrils, but some can be seen to transverse the fibrils. 102,000×

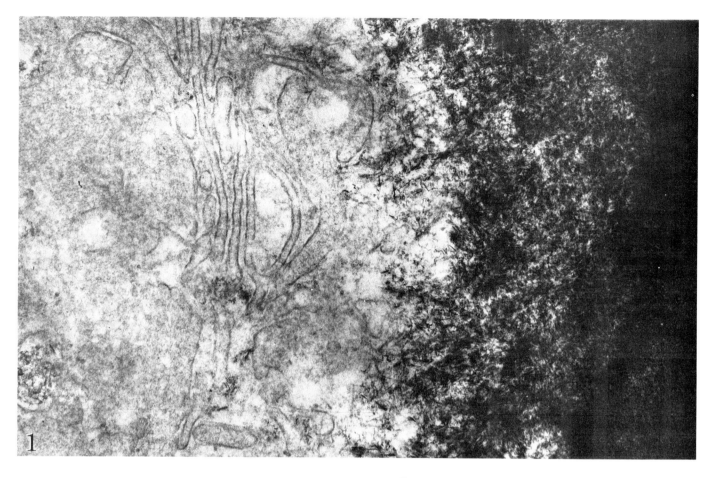

Bone—Osteoclast

Bone of endochondral or intramembranous origin is continually being rebuilt. The osteoclast, a large, multinucleated cell, is responsible for the dismantling of the bone.

1. The osteoclast has a markedly folded surface at the border adjacent to the bone. The bone itself has a peripheral area of rarefaction facing the osteoclast. Little evidence of fibrous stroma remains within the rarefied bone. Vacuoles with some mineral content are present in the osteoclast. 54,000 ×

Bone Marrow—Sternum

Bone marrow, where most of the blood cells are formed during adult life, is made up of a number of cells in various stages of development.

2. L. In sections stained with toluidine blue there is a great range in size of cells from that of the megakaryocyte (m) to the erythrocyte, in the nuclear shape and their chromatin distribution, in the density of the cytoplasm, and in the number and the size of cytoplasmic granules in individual cells. The most prominent cells of the bone marrow are the megakaryocytes. The nuclei of these extremely large cells are large, often multilobed or multiple. In the multinucleated stage, a more mature phase of the cell, the chromatin becomes condensed around the periphery of the nuclei. The cytoplasmic portion of the cell is generally extremely large with small, moderately dense granules more distinct in the mature cells. Among the megakaryocytes are many of the erythropoietic series of cells. Developing hemoglobin in nucleated erythrocytes (ne) appears as a moderately stained, homogeneous substance in the cytoplasm. Other cells of the granulocytic series are scattered throughout the field. Those with the dense cytoplasm are the heterophils in various stages of maturation. The cells in mitosis are evidence of the high rate of cell division in the bone marrow. 640 ×

3. Parts of a multilobed nucleus and of the cytoplasm of a megakaryocyte. Plasma membranes develop in the cytoplasm and eventually form complete partitions between the numerous future blood platelets or thrombocytes. Granules with a very densely stained material superimposed on a less dense background are present in most of the compartments of the cytoplasm. 6800 ×

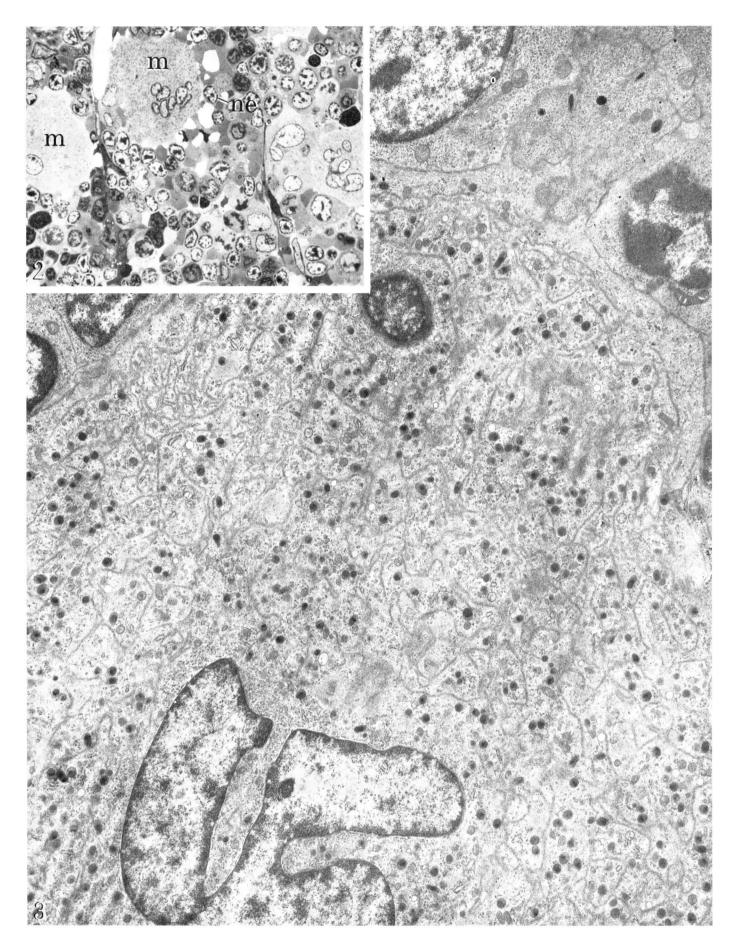

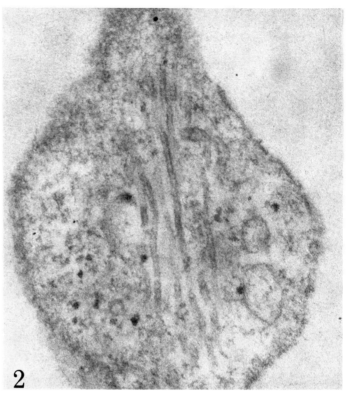

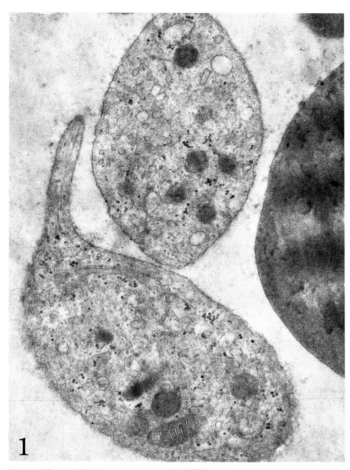

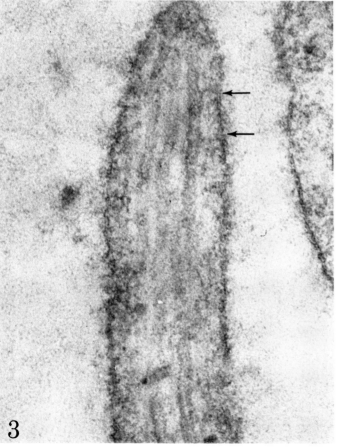

Bone Marrow and Platelet

1. The platelets, when released from the megakaryocyte of the bone marrow, escape into the bloodstream, where they appear as individual cytoplasmic units. The platelet contains dense granules, mitochondria, agranular endoplasmic reticulum in circular profiles and in tubular form, and cytoplasmic microtubules. The cytoplasmic microtubules form a circumferential band in the peripheral cytoplasm of the platelet. When the platelet is sectioned through its center, these microtubules appear as small, circular profiles at the two poles. In the platelet, in which a pseudopod-like extension has developed, the peripheral band of microtubules is drawn out to the tip of the process. 25,000 ×

2. A section through a peripheral segment of platelet demonstrates the band of slightly irregular microtubules with varying luminal density. 105,000 ×

3. The cytoplasmic microtubules, drawn out to the tip of the cytoplasmic process, appear to be connected to the plasma membrane by filamentous bridges (arrows). The surface of the platelet has a slightly coated appearance. 150,000 ×

4. As the megakaryocyte matures, the chromatin becomes more concentrated at the periphery of the nucleus. The majority of the cytoplasm has been lost as the platelets have been released into the blood stream. A number of dense granules are present in the one obvious remaining platelet and in the cytoplasm between the lobes of the nucleus. Immature cells with large nuclei, varying slightly in concentrations of chromatin and with pale cytoplasm, surround the megakaryocyte. 7600 ×

5. In the erythropoietic series of cells, the hemoglobin develops as a part of the cytoplasm. At this stage these cells are nucleated erythrocytes. A membranous partition develops between the hemoglobin-filled portion and the remainder of the cell. The hemoglobin with its surrounding plasma membrane then separates off from the remainder of the cell to become the anucleate erythrocytes of the circulating blood. 7400 ×

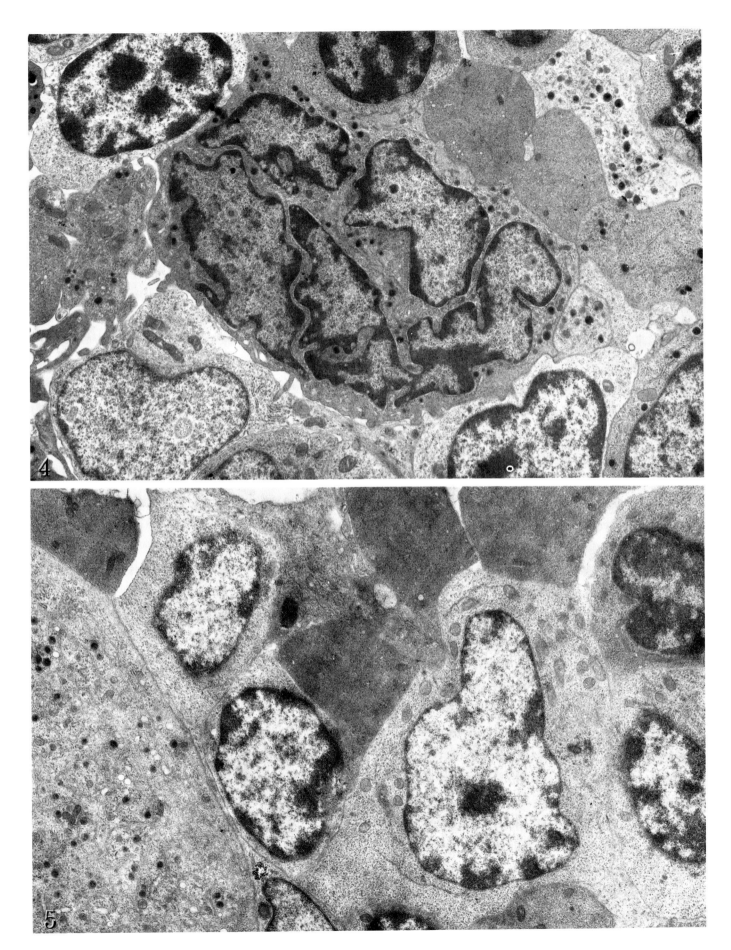

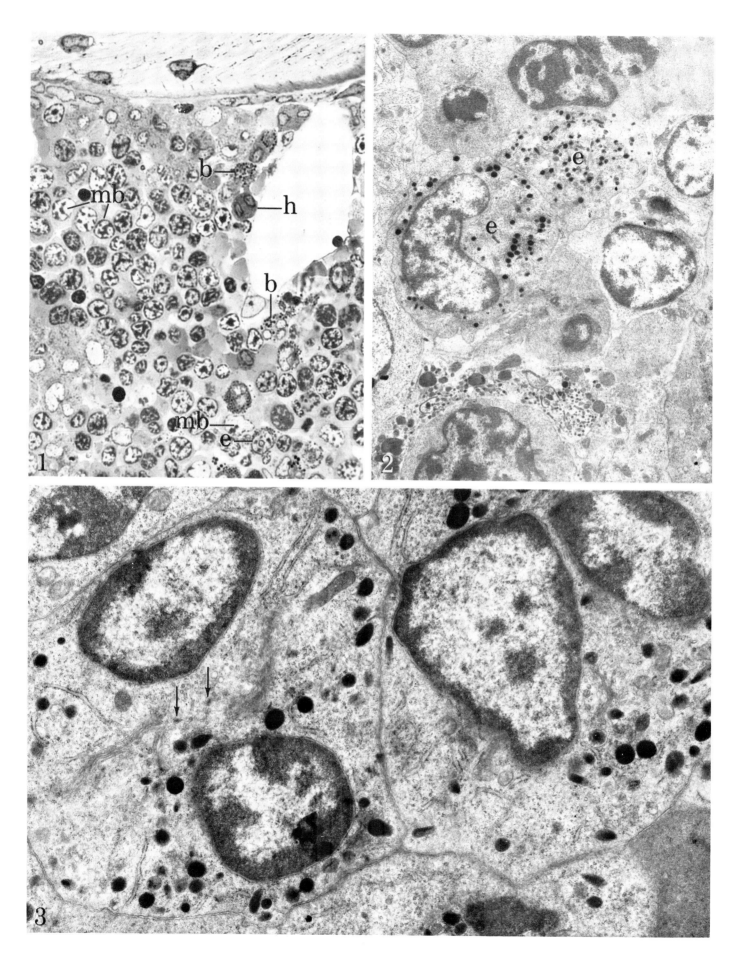

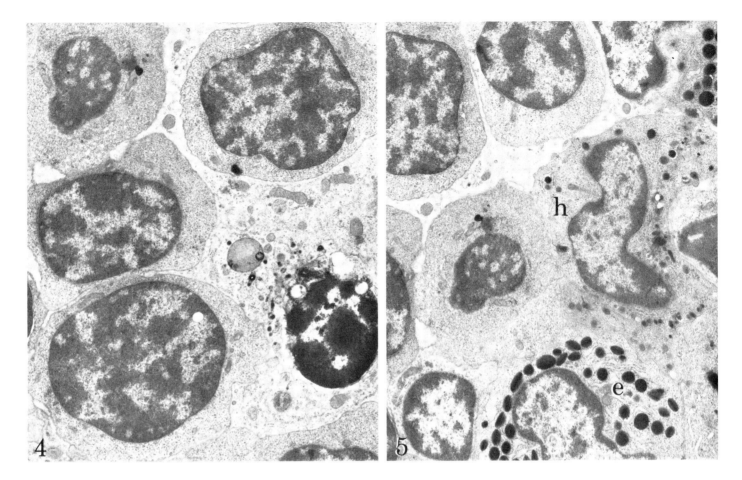

Bone Marrow—Sternum

The identification of the stages of development of some of the various types of cells in the bone marrow has been accomplished by exhaustive studies of nuclear and granule morphology and a correlation of light microscopic, histochemical, autoradiographic, and electron microscopic observations.

1. L. As a general rule, maturation of the cells is accompanied by a condensation of the chromatin and an adoption of the definitive form of the nucleus for that particular cell. The blast cells (mb) are nongranulated. Granules in immature cells, when they appear, are considerably larger than those of the mature cell. In the basophilic (b) myelocytes the nuclei become lobulated and the granules are dense against a pale background. In the eosinophilic myelocyte (e) the nuclei become crescent or doughnut shaped with maturity and the cytoplasm contains less distinct granules. The heterophilic myelocytes (h) have a dense cytoplasm, and the nucleus becomes quite lobulated with maturity. Granules are visible in immature or intermediate forms, but are difficult to resolve in light microscopy in mature forms. 720×

2. Intermediate eosinophilic myelocytes (e) contain a number of round to ovoid granules against a moderately pale background. Granules are present in the juxtanuclear Golgi zone. The nucleus in this section appears crescent shaped with chromatin only moderately stained. The numerous small to very large, less dense granules are in a cytoplasmic process of a macrophage. 5200×

3. Basophilic myelocytes contain a number of very dense granules in a pale cytoplasm. There is a relatively small amount of granular endoplasmic reticulum. Small granules of the very dense material occur in vesicles of the Golgi apparatus (arrows). In these relatively mature basophils the chromatin is mostly at the periphery of the bilobed nuclei. 15,600×

Bone Marrow

4. It is difficult to identify types of immature cells until granules make their appearance in the cytoplasm. These cells with the densely stained cytoplasm possibly represent immature cells of the erythropoietic series. A degenerated cell with a pyknotic, vesiculated nucleus is in the center of the field. 7400×

5. Immature and intermediate heterophylic myelocytes (h) are compared with an eosinophilic myelocyte (e) in which the central dense bands of the granules have become distinguishable. The background cytoplasm of the eosinophil is much less dense than in the heterophil. 6600×

REFERENCES

CONNECTIVE TISSUE

DuPraw, E. J. (1968). Physical chemistry of macromolecules. "Cell and Molecular Biology," pp. 287–343. Academic Press, New York.

Hall, D. A., ed. (1963, 1964, 1965). "International Review of Connective Tissue Research," Vols. 1, 2, and 3. Academic Press, New York.

Hodge, A. J. (1967). Structure at the electron microscopic level. *In* "Treatise on Collagen" (G. N. Ramachandran, ed.), Vol. 1, p. 185–205. Academic Press, New York.

Jackson, S. F. (1964). Connective tissue cells. *In* "The Cell" (J. Brachet and A. E. Mirsky, eds.), Vol. 6, pp. 387–520. Academic Press, New York.

Kobayashi, T., Midtgard, K., and Asboe-Hansen, G. (1968). Ultrastructure of human mast-cell granules. *J. Ultrastruct. Res.* **23**, 153–165.

Rhodin, J. A. G. (1967). Organization and ultrastructure of connective tissue. *In* "The Connective Tissue" (B. M. Wagner and D. E. Smith, eds.), pp. 1–16. Williams & Wilkins, Baltimore, Maryland.

Riley, J. F. (1959). "The Mast Cells." Livingstone, Edinburgh and London.

Ross, R. (1968). The connective tissue fiber forming cell. *In* "Treatise on Collagen" (B. S. Gould, ed.), Vol. 2, Part A, pp. 2–82. Academic Press, New York.

CARTILAGE

Anderson, H. C. (1967). Electron microscopic studies on induced cartilage development and calcification. *J. Cell. Biol.* **35**, 81–101.

Bonnuci, E. (1967). Fine structure of early cartilage calcification. *J. Ultrastruct. Res.* **20**, 33–50.

Greer, R. G., Brennan, W. T., and Mankin, H. J. (1967). Protein synthesis in epiphyseal cartilage. I. Incorporation rates and distribution of glycine-H3 and Na2S3404 *in vitro*. *Lab Invest.* **16**, 496–503.

Salpeter, M. M. (1968). H^3-proline incorporation into cartilage: Electron microscope autoradiographic observations. *J. Morphol.* **124**, 387–422.

BONE

Bélanger, L. F. (1967). Autoradiographic studies of the formation of the organic matrix of cartilage, bone and the tissue of teeth. *Ciba Found. Symp., Bone Struct. Metab.* pp. 75–87.

Glimcher, M. J., and Krane, S. M. (1968). The organization and structure of bone, and the mechanism of calcification. *In* "Treatise on Collagen" (B. S. Gould, ed.), Vol. 2, Part B, Academic Press, New York.

BONE MARROW

Ackerman, G. A. (1968). Ultrastructure and cytochemistry of the developing neutrophil. *Lab. Invest.* **19**, 290–302.

Bainton, D. F., and Farquhar, M. G. (1968b). Differences in enzyme content of azurophil and specific granules of polymorphonuclear leukocytes. II. Cytochemistry and electron microscopy of bone marrow cells. *J. Cell. Biol* **39**, 299–317.

Jensen, K. G., and Kilmann, S. A. (1968). "Blood Platelets. Structure, Formation and Function," Vol. 1, No. 2, Ser. Haemotol, Williams & Wilkins, Baltimore, Maryland.

Miller, F., De Harven, E., and Palade, G. E. (1966). The structure of eosinophil leukocyte granules in rodents and in man. *J. Cell Biol.* **31**, 349–362.

Sandborn, E. B., Lebuis, J. J., and Bois, P. (1966). Cytoplasmic microtubules in blood platelets. *Blood* **27**, 247–252.

Wetzel, B. K., Horn, R. G., and Spicer, S. S. (1967). Fine structural studies on the development of heterophil, eosinophil, and basophil granulocytes in rabbits. *Lab. Invest.* **16**, 349–382.

STRATIFIED
SQUAMOUS
EPITHELIUM

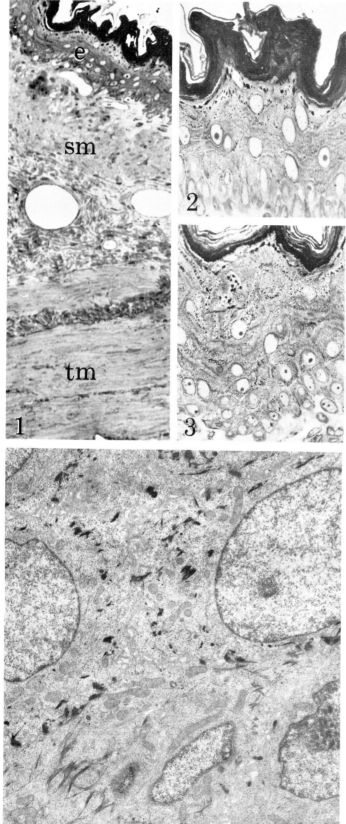

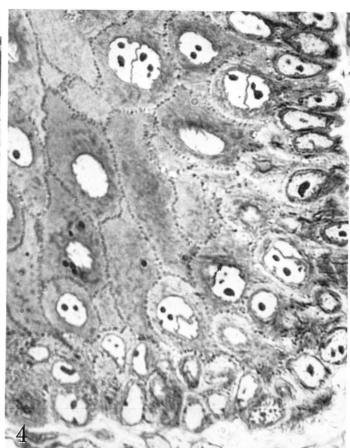

Stratified Squamous Epithelium

Stratified squamous epithelium is one of the most extensive of tissues. It covers the external surface of the body. It lines the mouth, the tongue, the epiglottis, and the esophagus in most animals and the fundus of the stomach as well in the rat.

1. L. Esophagus. The layers of the wall of the esophagus are: a tunica muscularis (tm) of longitudinally and circularly oriented smooth muscle; a submucosa (sm) of loose connective tissue, vessels, and nerves; a tunica mucosae consisting of a lamina muscularis mucosae, a lamina propria mucosae, and a stratified squamous epithelium (e) facing the lumen. The epithelium appears as a number of layers of cells, changing in character until they form a dense cornified surface layer. 170×

2. L. Esophagus. The layers of the epithelium consist of: the stratum Malpighii, which may be subdivided into, (a) the stratum germinativum or basale, the cuboidal, basal cells which stain deeply basophilic; (b) the stratum spinosum, the prickle cell layer consisting of several layers of polyhedral cells; the stratum granulosum, in which the nuclei become indistinct in the uppermost layers and which contains dense granules of keratohyalin and the dense stratum corneum. 440×

3. L. Cornified squamous epithelium extends into the fundus of the stomach in the rat. The cornified layer is thinner and less irregular than in the esophagus, but otherwise the layers are similar. 440×

4. L. Tongue. Numerous densities along the intercellular junctions outline the cell borders in stratified squamous epithelium. These densities are desmosomes. Many of the cells in this location of the stratum spinosum are binucleate with extremely prominent multiple nucleoli. 800×

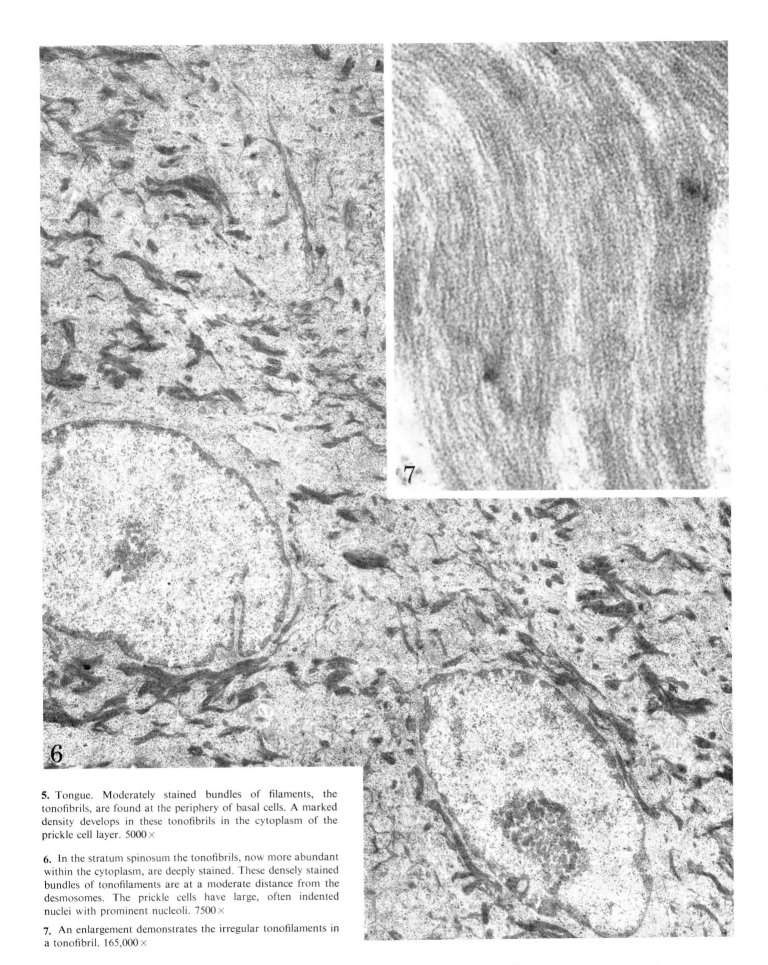

5. Tongue. Moderately stained bundles of filaments, the tonofibrils, are found at the periphery of basal cells. A marked density develops in these tonofibrils in the cytoplasm of the prickle cell layer. 5000×

6. In the stratum spinosum the tonofibrils, now more abundant within the cytoplasm, are deeply stained. These densely stained bundles of tonofilaments are at a moderate distance from the desmosomes. The prickle cells have large, often indented nuclei with prominent nucleoli. 7500×

7. An enlargement demonstrates the irregular tonofilaments in a tonofibril. 165,000×

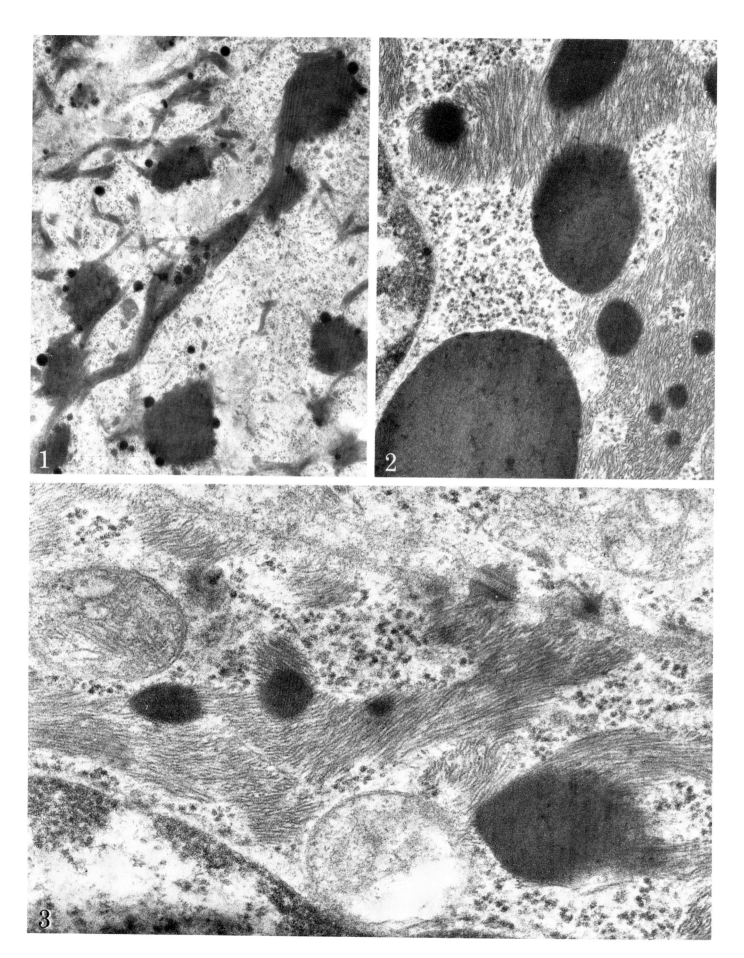

56   Stratified Squamous Epithelium

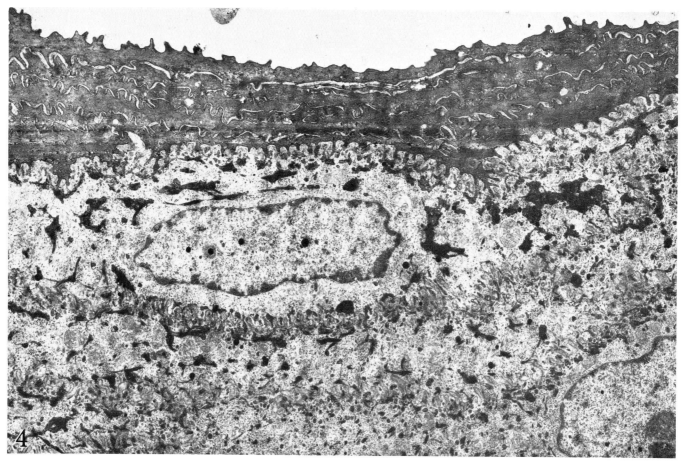

Skin

1. In the transition from stratum spinosum to stratum granulosum a dense homogeneous substance, keratohyaline, is deposited in the tonofibrils. Smaller denser circular granules are found between the bundles of tonofilaments. 27,000×

2. Some of these dense keratohyaline granules appear to be within the tonofibrils. 52,500×

3. The tonofilaments appear closely associated with the desmosomes. Filaments can be detected within the granules. 75,000×

4. With the death of the cell a remarkable change in appearance between the granular and the cornified layers develops. The cells are forced toward the surface by the continuous process of cell division in the deeper stratum germinativum. The stratum corneum represents a number of layers of individual dead cells wherein the nucleus and all semblance of organelles have disappeared. The deepest layer of cells in the stratum corneum is adherent to the most superficial layer of the granulosum, but the interspaces between layers increase progressively in size toward the surface. 6800×

5. In the most superficial layers of the stratum corneum only thin processes of neighboring cells make contact at degenerating desmosomes. The degenerated cells still contain a skeleton of filaments. 70,000×

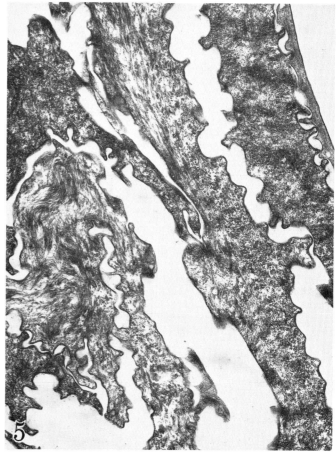

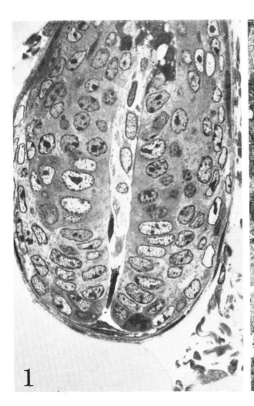

Hair Follicle

1. L. In the hair follicle a circular depression about this specialized papilla extends to a great depth. The bulb of the hair comprises a fold of epidermis around an elongated, vascularized papilla of the dermis. The hair follicle extends into the subcutaneous tissue. A thin fibrous coat, the connective tissue or dermal sheath, is carried down with the bulb. The stratum germinativum constitutes the outermost layer of epidermal cells of the bulb and those which envelop the dermal papilla as well. 900×

2. In the development of the hair, the cells of the internal root sheath and cuticle undergo a transformation by the proliferation of filaments followed by the deposition of tricohyalin in the form of large dense granules. This eventually involves the whole cell to such an extent that it becomes hardened and the nucleus becomes pyknotic. Here Huxley's layer demonstrates the large, dense tricohyalin granules while the cells of Henle's layer have undergone the hardening. The cytoplasm of the latter appears homogeneous and the nuclei are irregular in outline and are pyknotic. The cells of the external root sheath, which do not become involved in the hardening, show some vacuolation. At a more superficial level keratohyalin granules, typical of the epidermis develop in the cells of the external root sheath. The dermal of fibrous sheath consists of: an inner typical basal lamina of homogeneous material and a network of reticular filaments which has been termed the glassy membrane; a layer of moderately dense fibrous tissue with a number of fibroblasts; and an external layer of longitudinally oriented, ill-defined collagen fibers, which are seen here in transverse section. 2250×

The Sebaceous Gland

3. L. The sebaceous gland, a derivative of the epidermal layer, discharges its secretion into the neck of the hair follicle. There is a peripheral single layer of flattened cells continuous with the basal layer of the epidermis. The cells at the depths of the gland are large, pale, and homogeneous. In succeeding layers, the cells, neatly layered one upon the other, develop an increasing number of clear vesicles. The nuclei are large with multiple, prominent nucleoli in the depth of the gland. In the more superficial half of the gland, the nuclei become gradually more pyknotic until, in the outermost cells, they are degenerated into irregular dark masses. Disintegrated parts of the most superficial cell are discharged into the follicular canal and form a secretion about the base of the hair. 750×

4. The peripheral or basal layer consists of flattened cells with small, dense nuclei. In other cells, nearing the apex of the gland, the cytoplasm contains numerous clear vacuoles and small, intervening, moderately dense granules. The chromatin has become concentrated into dense masses throughout the nucleus. 5700×

The Eccrine Sweat Gland

5. L. The coiled, secretory portion of this unbranched, tubular gland is lined by an epithelium whose cells fall into two categories: a lighter cell, which has a low to moderate content of supranuclear secretory material, and a darker, taller cell, in which the apex is filled with more dense material. In these animals both types appear to be associated with intercellular canaliculi (arrows). Myoepithelial cells are located at the base of the glandular epithelium. 750×

6. The supranuclear portion of the darker cell is packed with irregular dense granules. There is an abundant granular endoplasmic reticulum and prominent multiple Golgi zones. 10,600×

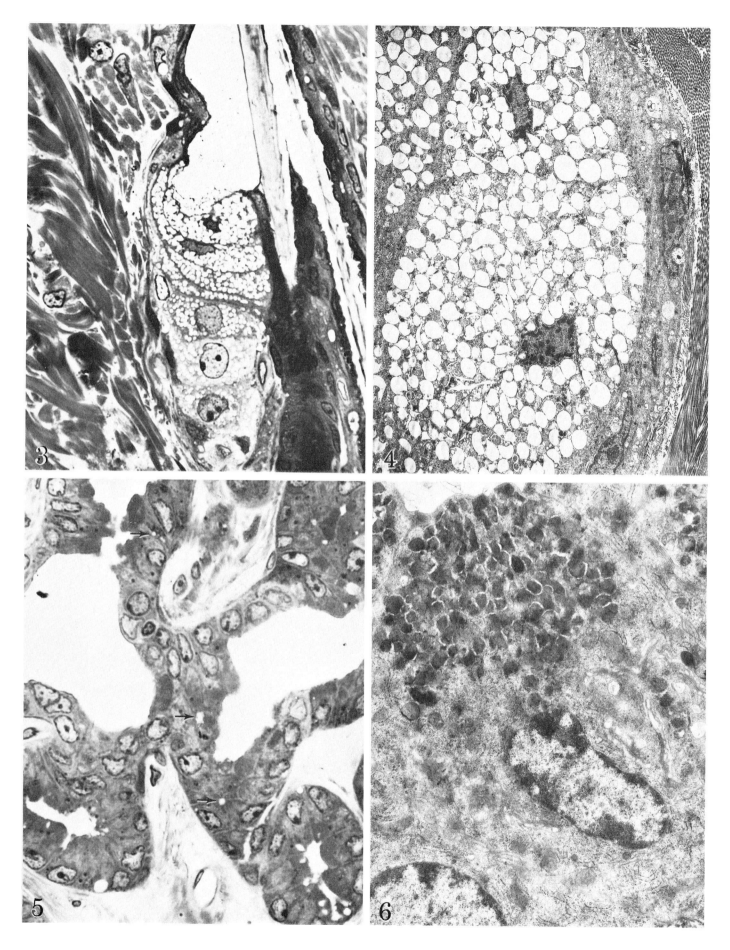

REFERENCES

Allen, A. C. (1967). "The Skin," 2nd ed. Grune & Stratton, New York.

Bourrlond, A. (1968). "L'innervation cutanée. Recherches sur son ultrastructure et analyse des techniques de microscopie optique." Masson, Paris.

Brody, I., and Larsson, K. S. (1965). Morphology of mammalian skin: Embryonic development of the epidermal sub-layers. *In* "Biology of the Skin and Hair Growth" (A. G. Lyne, and B. F. Short, eds.), pp. 267–290. Elsevier, Amsterdam.

Dalton, A. J., and Felix, M. D. (1964). Cytologic and cyto-chemical characteristics of the Golgi substance of epithelial cells of the epididymis — *in situ*, in homogenates and after isolation. *Am. J. Anat.* **94**, 171–208.

Ellis, R. A. (1965). Fine structure of the myoepithelium of the eccrine sweat glands of man. *J. Cell Biol.* **27**, 551–563.

Farquhar, M. G., and Palade, G. E. (1965). Cell junctions in amphibian skin. *J. Cell Biol.* **26**, 263–292.

Kurosumi, K. (1968). Skin. *In* "Fine Structure of Cells and Tissues. Electron Microscopic Atlas" (K. Kurosumi, and H. Fujita eds.), Vol. 5, pp. 183–277. Igaku Shoin, Tokyo.

Leblond, C. P., Greulich, R. C., and Pereira, J. P. M. (1964). Relationship of cell formation and cell migration in the renewal of stratified squamous epithelia. *Advan. Biol. Skin* **5**, 39–67.

Matoltsy, A. G. (1966). Membrane-coating granules of the epidermis. *J. Ultrastruct. Res.* **15**, 510–515.

Mishima, Y. (1966). Melanosomes in phagocytic vacuoles in Langerhans cells. *J. Cell Biol.* **30**, 417–423.

Montagna, W. (1962). "The Structure and Function of Skin," 2nd ed. (1st ed., 1956). Academic Press, New York.

Montagna, W., and Lobitz, W. C., Jr., eds. (1964). "The Epidermis." Academic Press, New York.

Parakkal, P. F. (1966). The fine structure of the dermal papilla of the guinea pig hair follicle. *J. Ultrastruct. Res.* **14**, 133–142.

Parakkal, P. F., and Matoltsy, A. G. (1968). An electron microscopic study of developing chick skin. *J. Ultrastruct. Res.* **23**, 403–416.

Roth, S. I., and Jones, W. A. (1967). The ultrastructure and enzymatic activity of the boa constrictor skin during the resting phase. *J. Ultrastruct. Res.* **18**, 304–323.

Snell, R. S. (1965). An electron microscopic study of keratin-ization in the epidermal cells of the guinea-pig. *Z. Zellforsch. Mikroskop Anat.* **65**, 829–846.

Szabo, G. (1965). Current state of pigment research with special reference to the macromolecular aspects. *In* "Biology of the Skin and Hair Growth" (A. G. Lyne, and B. F. Short, eds.), pp. 705–726. Angus & Robertson, Sydney, Australia.

Winkelmann, R. K. (1967). Normal structure of the skin. *In* "Dermatopathology" (H. Montgomery, ed.), Vol. 1, pp. 23–53. Harper (Hoeber), New York.

Wolff, K. (1967). The fine structure of the Langerhans cell granule. *J. Cell Biol.* **34**, 468–473.

Zelickson, A. S. (1967). "Ultrastructure of Normal and Abnormal Skin." Lea & Febiger, Philadelphia, Pennsylvania.

THE NERVOUS
SYSTEM

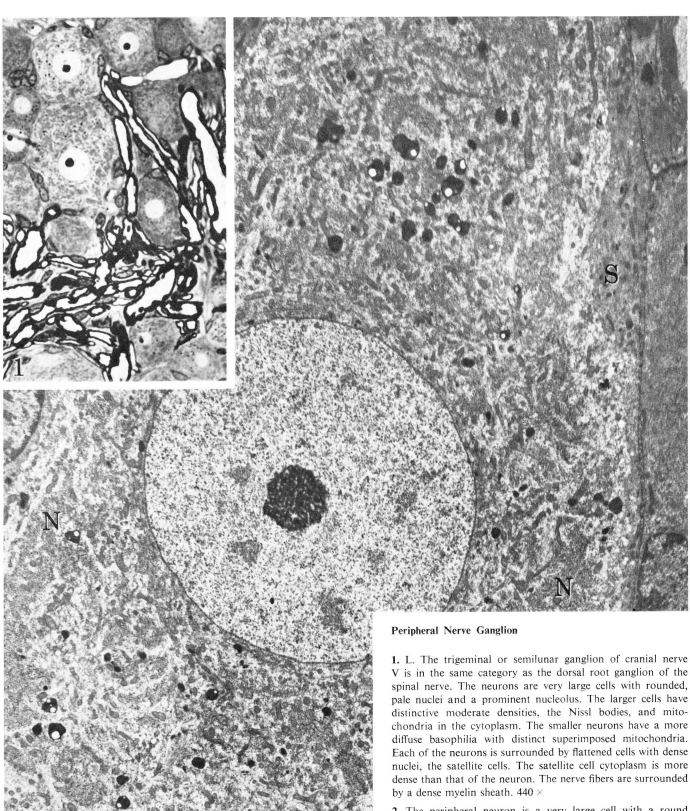

Peripheral Nerve Ganglion

1. L. The trigeminal or semilunar ganglion of cranial nerve V is in the same category as the dorsal root ganglion of the spinal nerve. The neurons are very large cells with rounded, pale nuclei and a prominent nucleolus. The larger cells have distinctive moderate densities, the Nissl bodies, and mitochondria in the cytoplasm. The smaller neurons have a more diffuse basophilia with distinct superimposed mitochondria. Each of the neurons is surrounded by flattened cells with dense nuclei, the satellite cells. The satellite cell cytoplasm is more dense than that of the neuron. The nerve fibers are surrounded by a dense myelin sheath. 440 ×

2. The peripheral neuron is a very large cell with a round nucleus and prominent nucleolus. The chromatin is finely granular and widely scattered in the large nucleus. The Nissl bodies (N), scattered through the cytoplasm of the perikaryon, have been shown to be accumulations of granular endoplasmic reticulum or ergastoplasm. In the young adult animal a considerable number of dense granules are present. These increase in both number and size with age. A thin border of denser satellite cell (S) cytoplasm encompasses the neuron. 4500 ×

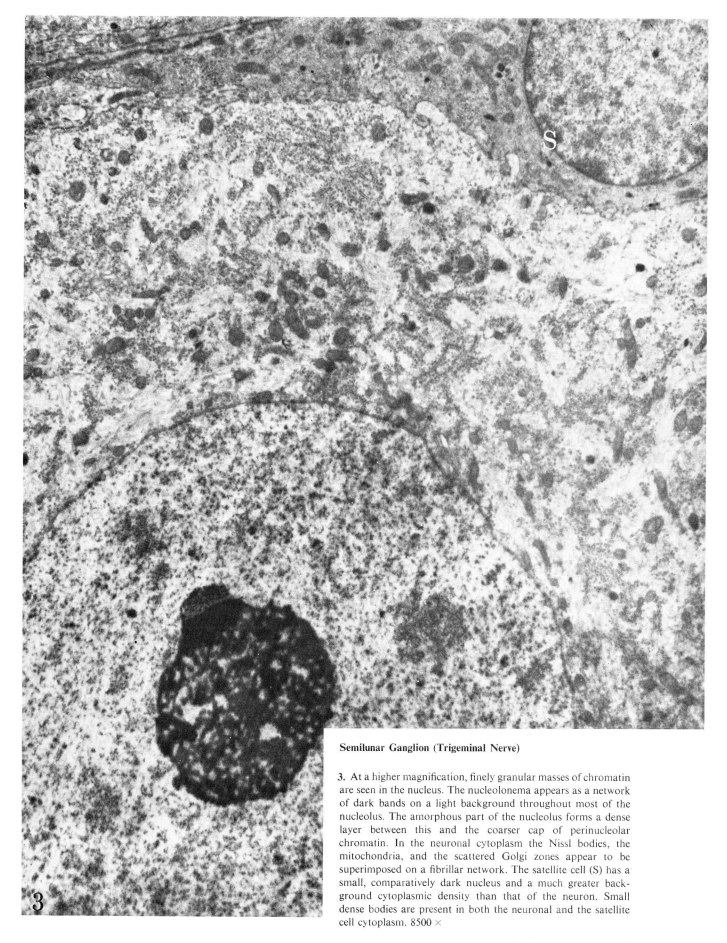

Semilunar Ganglion (Trigeminal Nerve)

3. At a higher magnification, finely granular masses of chromatin are seen in the nucleus. The nucleolonema appears as a network of dark bands on a light background throughout most of the nucleolus. The amorphous part of the nucleolus forms a dense layer between this and the coarser cap of perinucleolar chromatin. In the neuronal cytoplasm the Nissl bodies, the mitochondria, and the scattered Golgi zones appear to be superimposed on a fibrillar network. The satellite cell (S) has a small, comparatively dark nucleus and a much greater background cytoplasmic density than that of the neuron. Small dense bodies are present in both the neuronal and the satellite cell cytoplasm. 8500 ×

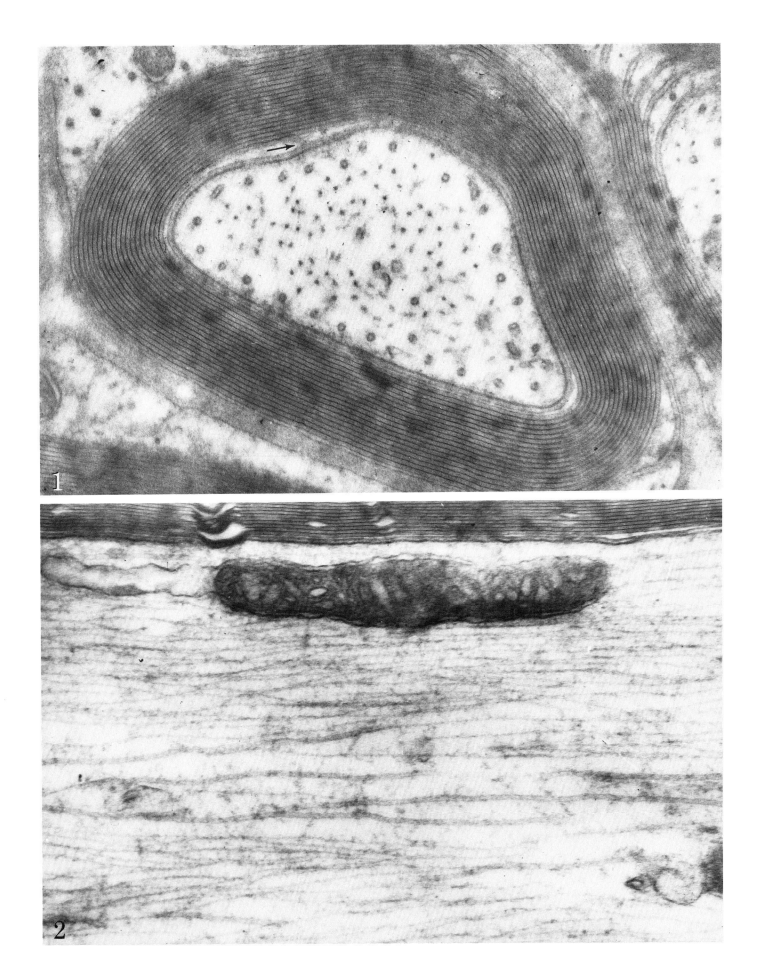

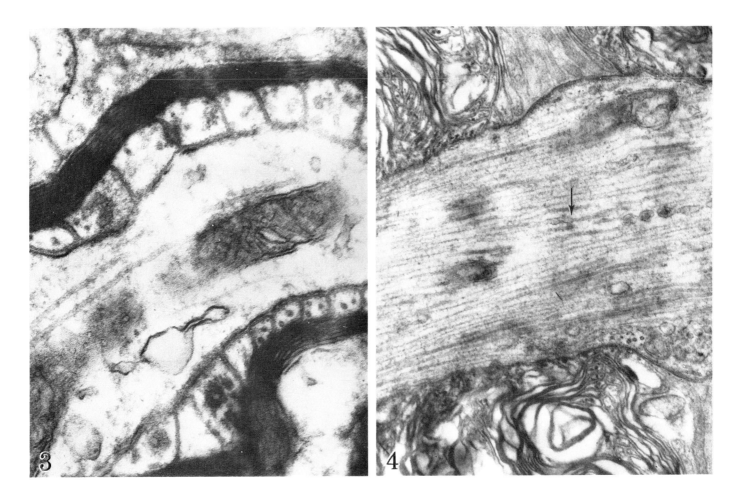

Myelinated Axons—Semilunar (Trigeminal) Ganglion

1. The myelin sheath of the axon is made up of many layers of Schwann cell plasma membrane. The myelin sheath, as a consequence of its ordered arrangement, has been employed extensively for the study of membranes. It has been observed that the inner dense layers of two adjacent membranes fuse to form the major periodicity of the sheath. The outer layers of adjacent membranes are also closely applied, but apparently not as securely fused to one another. These outer layers can be seen as faint interperiod lines. The myelin sheath formation begins as an embedding of an axon into a trenchlike depression or infolding of the surface of the Schwann cell. One wall of the Schwann cell surface which borders the infolding has advanced to envelop the axon. The head or crest of this advancing border (arrow) with a small amount of contained cytoplasm indicates its position after having wrapped the axon in several layers of membranes. The axon is bounded by its own plasma membrane. Cytoplasmic microtubules are sectioned transversely in both the axon and the Schwann cell. In addition a larger tube of the endoplasmic reticulum and a number of filaments have been sectioned transversely in the axon. There is no Golgi complex nor are there ribosomes in these axons. 110,000 ×

2. The cytoplasmic microtubules and the filaments stretch for great distances in the axon. The endoplasmic reticulum appears to be in continuity with the outer mitochondrial membrane. It has been reported that cytoplasmic microtubules give rise to the filaments in nerve fibers. 70,000 ×

Peripheral Nerve, Semilunar (Trigeminal) Ganglion

3. In the formation of a segment of myelin sheath by a single Schwann cell, the borders of the fold have advanced in steps along the axon as each lamina of two layers of the plasma membrane was laid down. As the fold was wrapped around the axon, a coil of cytoplasm was trapped in each border. In this micrograph one of these coils has been sectioned repeatedly as it encircles the axon. Cytoplasmic microtubules, included in this border, are seen between membrane infoldings which delimit the turns of the coil. For each turn the progression of the sheath along the axon is measured by twice the distance between adjacent loops of the coil since this same phenomenon has occurred at both extremities of the cytoplasmic fold of the Schwann cell. Finally, the extremities of the coils of the tube become continuous with the main body of the Schwann cell cytoplasm at the nodes of Ranvier. In this micrograph the addition of seven or eight periodic lines to the thickness of the sheath can be distinguished. This represents the addition of fourteen to sixteen layers of membrane. 70,000 ×

4. Node of Ranvier. The segment of the axon within the node is moderately dilated. Several layers of cytoplasmic folds from the borders of the Schwann cells contact the external surface of the axon. Cytoplasmic microtubules traverse the node. Segments of a number of the microtubules have a dense content and some appear to have dilated segments (arrow). A few cytoplasmic microtubules are sectioned transversely in the axon. 42,000 ×

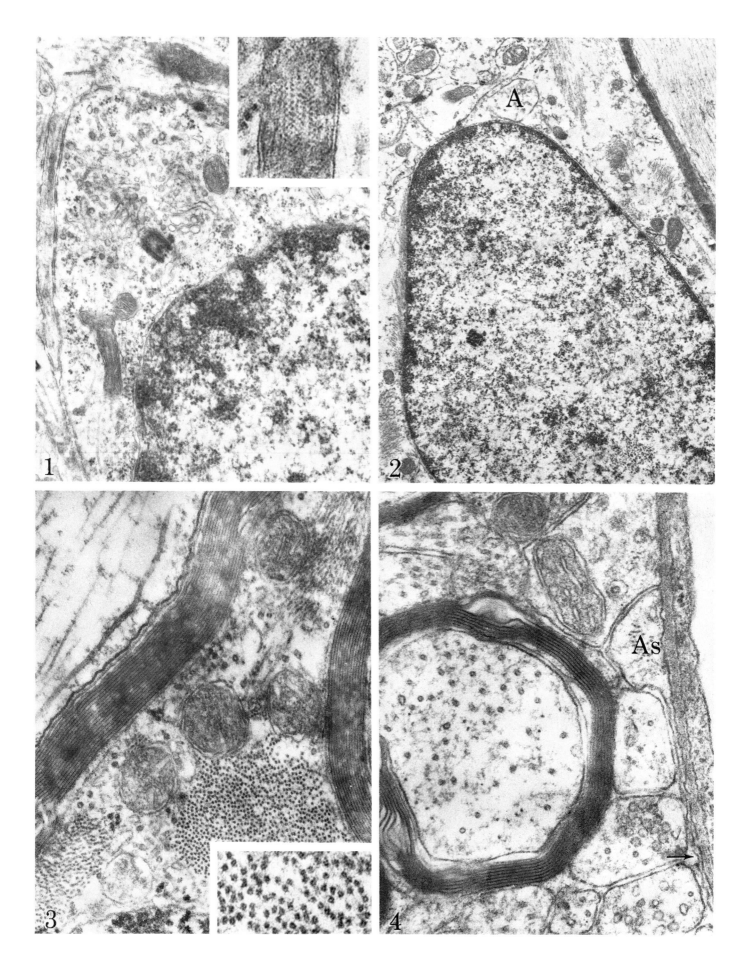

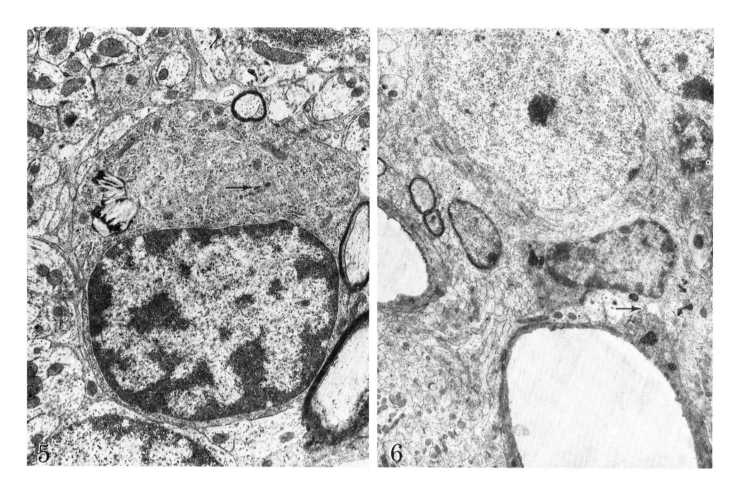

Neuroglial Cells

1. The astrocytes are a group of supportive glial cells with abundant pale cytoplasm and long processes which often end in expansions on capillaries. Some protoplasmic astrocytes have mutiple, well developed Golgi areas. Many cytoplasmic microtubules or filaments, centrioles, ribosomes, and glycogen granules are features of these cells and their processes. Bergmann astrocytes or Golgi epithelial cells may be included in this category. 27,500 ×

Inset: Mitochondria in astrocyte processes contain, on occasion, an internal substructure which has the appearance of helicoidal filaments. 122,000 ×

2. Fibrous astrocytes contain a higher proportion of filaments in their cytoplasm. This increase in filamentous content would appear to develop with the maturation of the astrocyte. Close to one border of the nucleus there is a band of massed cytoplasmic filaments which might be considered as tonofilaments. Similar tonofilaments extend into the cytoplasmic process which is stretched out along a myelinated fiber. A small axon (A) is included in a deep fold in the surface of the astrocyte. This is similar to that seen in the case of the Schwann cell. 12,000 ×

3. The filaments in one of these masses in the cytoplasm of an astrocyte have been sectioned transversely and others obliquely. Several cytoplasmic microtubules and mitochondria appear in transverse section. The filaments and the cytoplasmic microtubules are similar in dimension to those in the adjacent myelinated axon. A vesicle with lightly stained content is located near the nucleus of the astrocyte. 80,000 ×

Inset: At a greater enlargement the transversely sectioned filaments have a central clear zone surrounded by a densely stained periphery. 225,000 ×

4. Astrocyte processes (As), containing transversely sectioned filaments, intervene between a capillary and nerve processes, including a myelinated axon and many smaller dendrites, axones, and their endings. One astrocytic process appears to expand along the capillary as little more than an extended fold of its plasma membrane (arrow). Basal laminae are located along both the endothelium and the astrocytic process. 75,000×

5. The oligodendrocyte has a moderately densely stained cytoplasm which contains many organelles. In this cell dense granules have an unusual appearance with sharply demarcated light and dark zones. Smaller dense bodies are located in the Golgi apparatus (arrow). A small myelinated nerve fiber is embedded in the surface of the oligodendrocyte in the same manner as the peripheral nerve fibers are enveloped by the Schwann cell cytoplasm. 10,000 ×

6. A cell with moderately dense nuclear chromatin and cytoplasm is located between a stellate neuron and a capillary. The cytoplasm contains irregular-shaped, dense bodies. Similar dark cytoplasm, in which dense bodies are located, abuts on the capillary. It has been reported that cells with small dense nuclei and dense cytoplasm may be microglia. However these may be pericytes which have elongated processes leading to their pericapillary expanded base. A narrow isthmus of cytoplasm (arrow) interconnects the cell body and the pericapillary expansion in this instance. 6500 ×

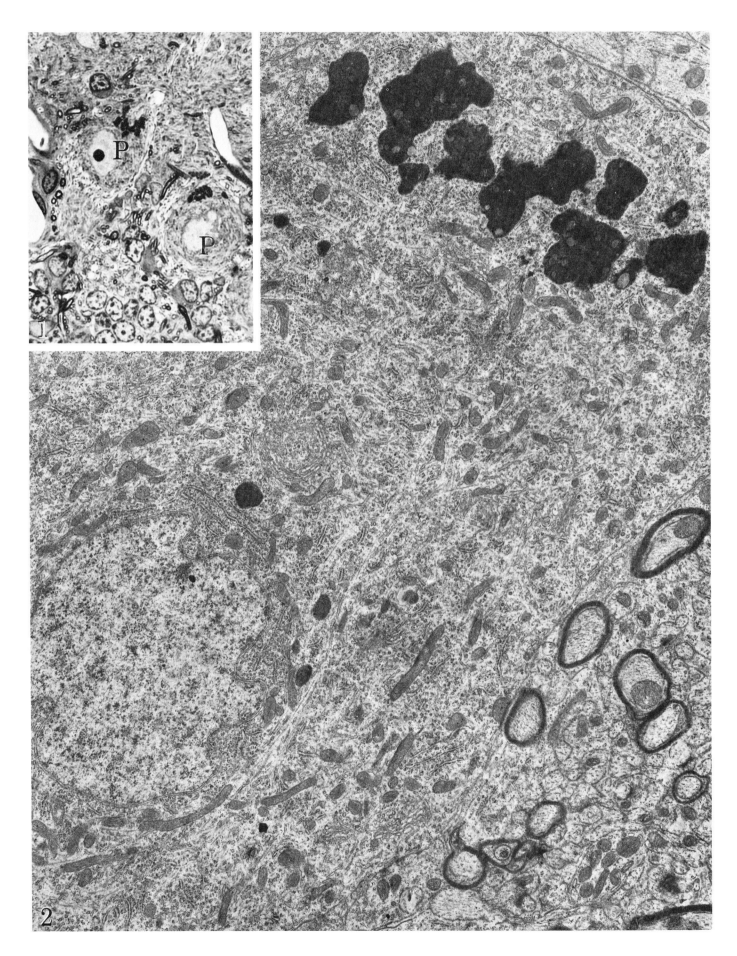

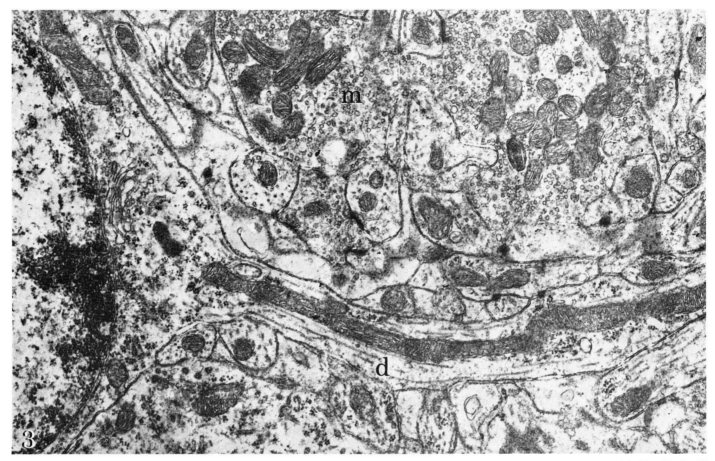

Cerebellum—Aged Rat

1. L. In the aged animal there is a marked increase in the number of dense bodies in the Purkinje cell (P). Accumulations of lipofuscin granules are concentrated at the base of the dendrites in the perikaryon. Smaller lysosomal-dense granules can be seen around the large, pale nuclei. Smaller cells with dark cytoplasm are oligodendrogliocytes. Some of these have small, dense bodies in their cytoplasm. 720 ×

2. The abundant cytoplasm of the Purkinje cell contains numerous collections of granular endoplasmic reticulum, the Nissl bodies, and multiple Golgi zones. The lipofuscin granules in this 24-month-old rat are large and massed near the cell periphery at the base of the dendrites. The dense bodies near the Golgi complexes have all of the features of lysosomes. Adjacent to the Purkinje cell are numerous very small, un-myelinated and a few larger, myelinated nerve fibers of the molecular layer of the cerebellum. 10,500 ×

Nerve Fibers—Cerebellum

3. The dendrite (d) of a neuron in the granular layer contains a number of long mitochondria, slightly irregular cytoplasmic microtubules, endoplasmic reticulum, and free ribosomes. Numerous small axonal and dendritic nerve processes and paler astrocyte cytoplasmic processes intervene between this dendrite and a large mossy fiber terminal (m) which is filled with synaptic vesicles and mitochondria. Ribosomes, as a general rule, are not found in axons. 25,000 ×

4. Multiple synaptic plaques are found between large axon terminals and dendrites. Subsurface cysternae of the agranular reticulum are occasionally found in the dendrites. 97,000 ×

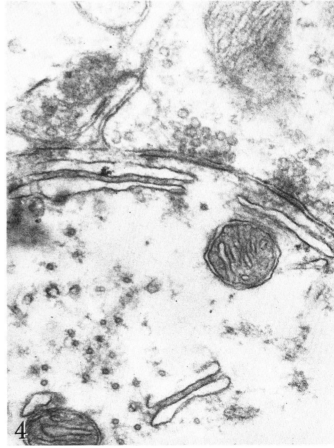

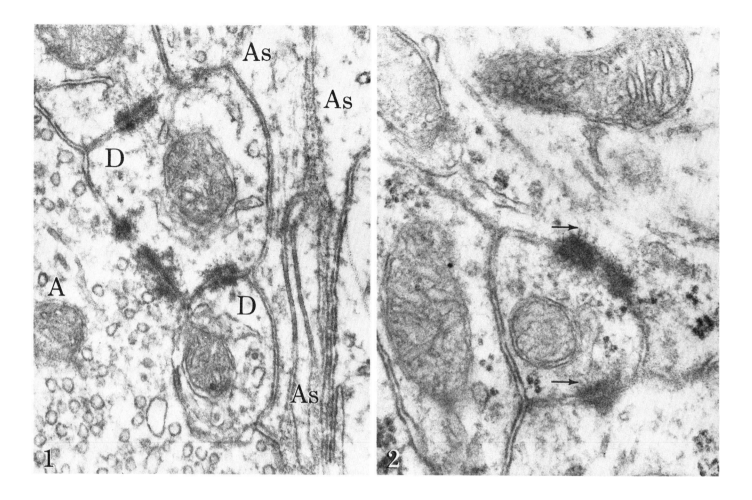

Neuronal and Glial Cell Junctional Complexes—Cerebellum

1. Various types of junctional complexes are found between cytoplasmic processes in the granular layer of the cerebellum. Axodendritic and dendrodendritic synapses are seen between nerve processes (A and D), while tight junctions (zonulae occludentes) occur between the cytoplasmic processes of the adjacent astrocytes (As). Many synaptic vesicles in the axonal terminals have surface filamentous stems. Some of the vesicles appear to be interconnected by these filaments. 116,000 ×

2. Curved filaments, some closely associated with a cytoplasmic microtubule, are embedded in the dense plaques of the dendrodendritic synapses (arrows). 94,000 ×

Nerve Endings and Synapses

3. Dendrodendritic synapses resemble desmosomes in their dimension, in the demonstrable central intercellular line, and in a series of lines resembling filaments which cross the intercellular space in other views. The dense cytoplasmic plaque is approximately equal in each of the dendrites involved in the junctional complex. 100,000 ×

4. Type-I axodendritic synapses have an intercellular space of approximately 300 Å as compared to approximately 200 Å in the adjacent zones. The synaptic vesicles are mostly circular with a diameter in a limited range of from 400 to 600 Å. Filaments appear to interconnect synaptic vesicles with the axonal membrane. The dense plaque is mostly confined to the dendritic aspect of this axodendritic synapse. 135,000 ×

5. Synaptic vesicles in the axonal terminal are often found in chains. The vesicles seem to be interconnected by filaments. Surface views of membranes appear as diffuse zones of only moderate density with fine lines in parallel (arrow). 120,000 ×

6. A synapse between an axon (A) and a dendrite (D) has been sectioned obliquely. Surface views of the synapse reveal lines in parallel in several locations (arrows). These lines might support the view that a linearly arranged substructure is present in the membrane. 200,000 ×

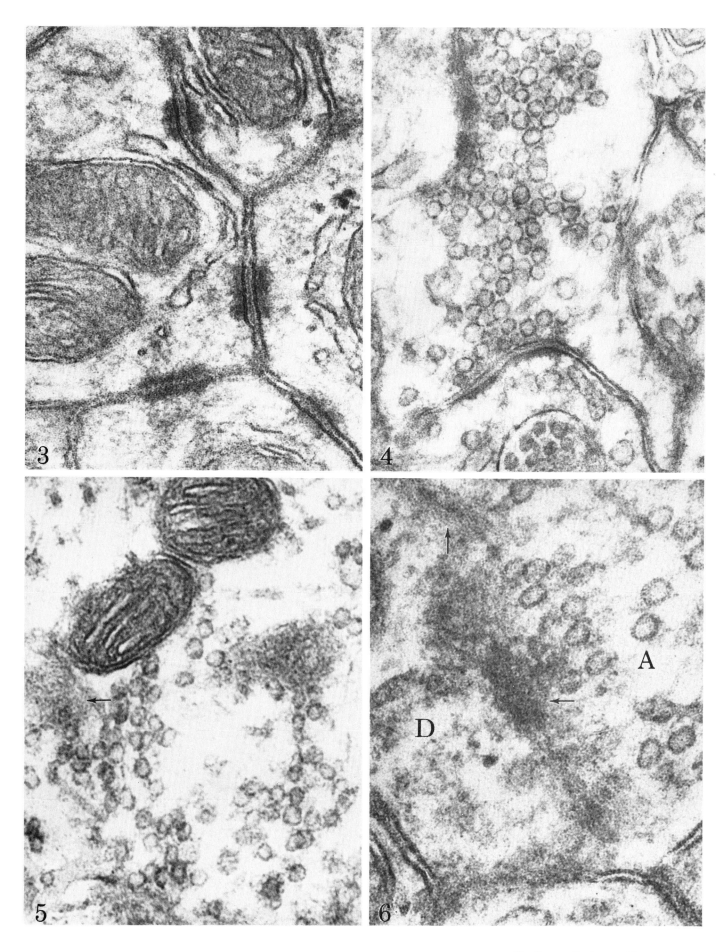

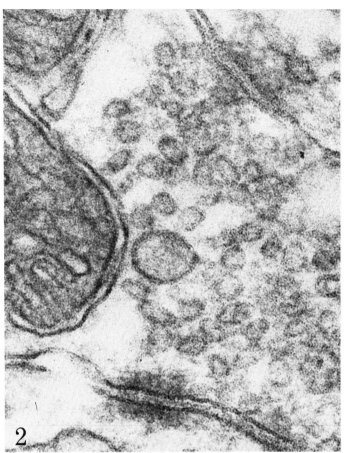

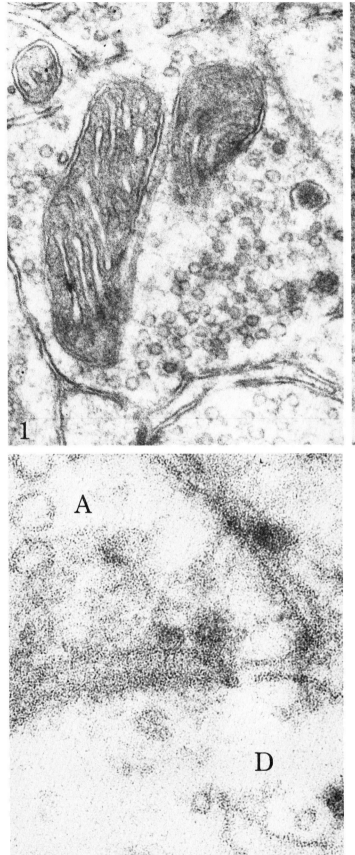

Nerve Endings and Synapses—Cerebellum

1. In unusual instances larger vesicles with dense content are found in the axonal terminals. Dense content is seen in some of the smaller vesicles, one of which has a surface coated with protruding filaments. A thin astrocyte process intervenes between the two axonal terminals. 100,000 ×

2. In the type-II axodendritic synapses there is little if any increase in width between the opposing plasma membranes of the synapse. A central dense line is found in each synaptic cleft. There is little or no difference in the density of the plaque on the axonal or the dendritic aspects of the synapse. Some of the synaptic vesicles in the axon terminal are slightly oval. There is one very large vesicle in the axon. A line of moderate density is present between the outer and the inner mitochondrial membranes. 200,000 ×

3. In a type-I axodendritic synapse filamentous-appearing bridges cross the intercellular space between the axon (A), and the dendrite (D). 255,000 ×

The Axodendritic Synapse—Type I

4. The axodendritic type-I synapse is an intercellular junctional complex wherein the intercellular space increases slightly within a limited zone. The intracytoplasmic dense plaque in the dendrite is much more distinct than in the axon. The membranes of the synaptic vesicles appear trilaminar or globular (g), but exhibit lines in parallel when sectioned obliquely or tangentially (arrows). 280,000 ×

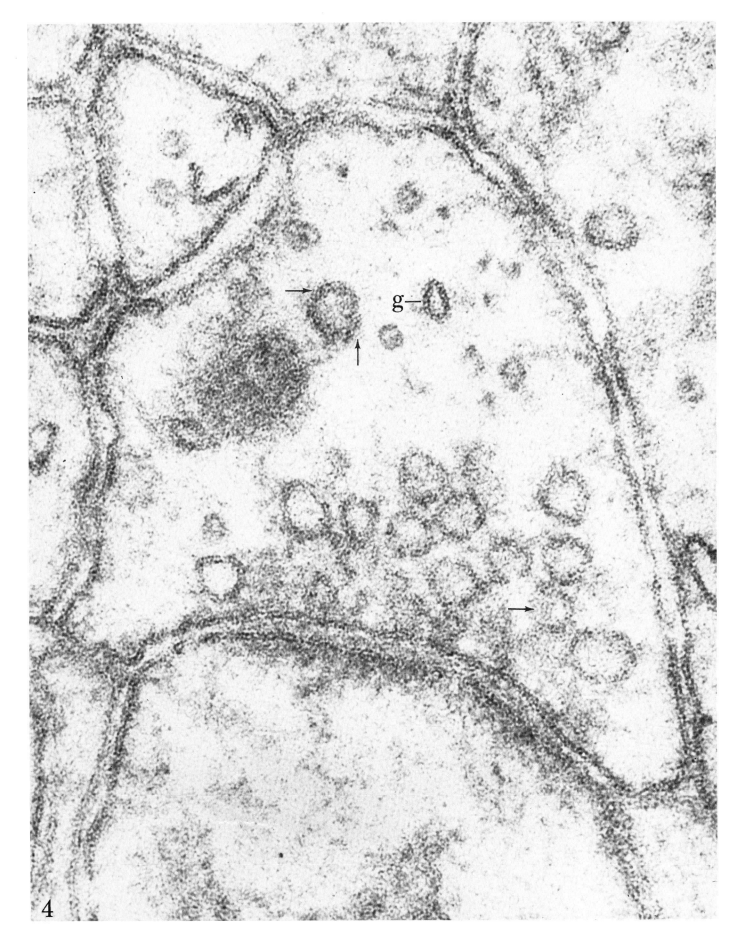

4

REFERENCES

Bodian, D. (1964). An electron microscopic study of the monkey spinal cord. *Bull. Johns Hopkins Hosp.* **114**, 13–19.

Cravioto, H. (1965). The role of Schwann cells in the development of human peripheral nerves. An electron microscopic study. *J. Ultrastruct. Res.* **12**, 634–651.

De Robertis, E. (1966). Synaptic complexes and synaptic vesicles as structural and biochemical units of the central nervous system. *In* "Nerve as a Tissue" (K. Rodahl and B. Issekutz, eds.), pp. 88–115. Harper (Hoeber), New York.

Eccles, J. C., Ito, M., and Szentagothai, J. (1967). "The Cerebellum as a Neuronal Machine." Springer, Berlin.

Fox, C. A., Hillman, D. H., Siegesmund, K. A., and Dutta, C. R. The primate cerebellar cortex: A Golgi and electron microscopic study. *In* "The Cerebellum" *Progr. Brain Res.* **25**, pp. 174–225 (C. A. Fox and R. S. Snider, eds.), Elsevier, Amsterdam.

Geren, B. B. (1954). The formation from the Schwann cell surface of myelin in the peripheral nerves of chick embryos. *Exptl. Cell Res.* **7**, 558–562.

Gray, E. G. (1964). Tissue of the central nervous system. *In* "Electron Microscopic Anatomy" (S. M. Kurtz, ed.), pp. 369–417. Academic Press, New York.

Kuhlenbeck, H., ed. (1967). "The Central Nervous System of Vertebrates," 5 vol. Academic Press, New York.

Ochs, S., Sabri, M. I., and Johnson, J. (1969). Fast transport system of materials in mammalian nerve fibers. *Science* **163**.

Palay, S. L. (1966). The role of neuroglia in the organization of the central nervous system. *In* "Nerve as a Tissue" (K. Rodahl and B. Issekutz, eds.), pp. 3–10. Harper (Hoeber), New York.

Palay, S. L., and Palade, G. E. (1955). The fine structure of neurons. *J. Biophys. Biochem. Cytol.* **1**, 69–88.

Pappas, G. D., (1966). Electron microscopy of neuronal junctions involved in transmission in the central nervous system. *In* "Nerve as a Tissue" (K. Rodahl and B. Issekutz, eds.), pp. 49–87. Harper (Hoeber), New York.

Weiss, P., Taylor, C., and Pillai, P. (1962). The nerve fiber as a system in continuous flow: Microcinematographic and electronmicroscopic demonstrations. *Science* **136**, 330.

Whittaker, V. P. (1959). The isolation and characterization of acetylcholine containing particles from brain. *Biochem. J.* **72**, 694–706.

Whittaker, V. P., and Gray, E. G. (1962). The synapse: Biology and morphology, *Brit. Med. Bull.* **18**, 223–228.

MUSCLE

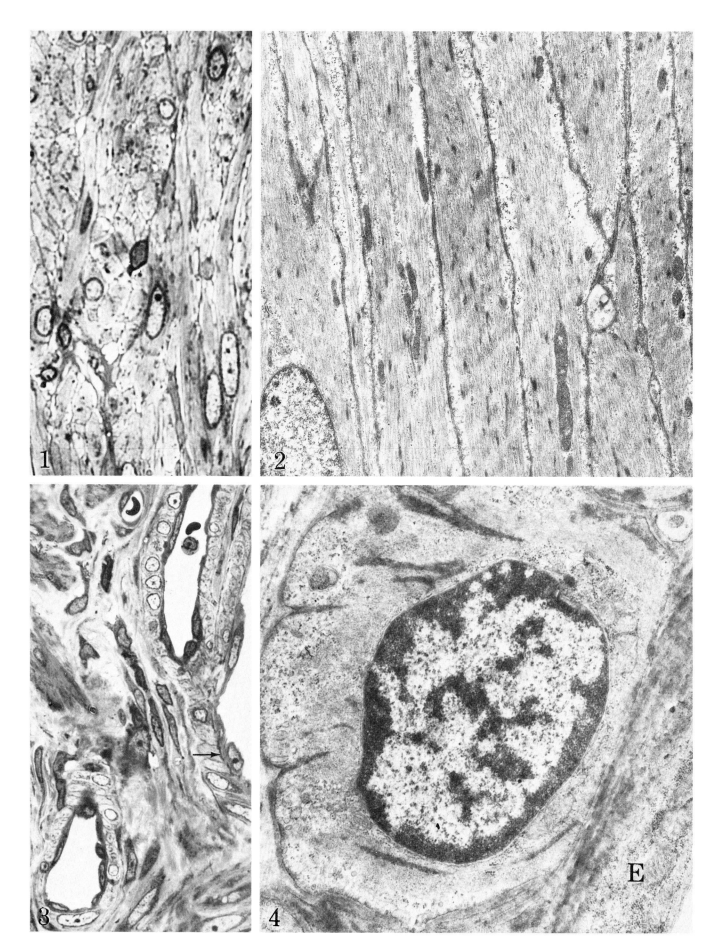

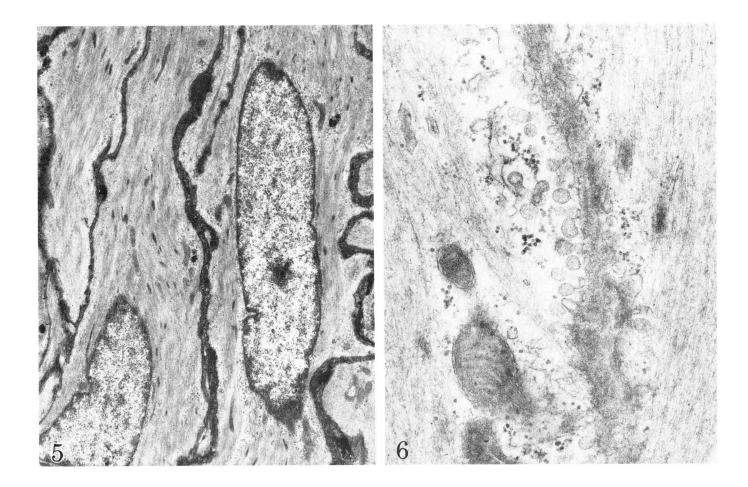

Smooth Muscle

1. L. Cells or fibers from alternate layers of smooth muscle in the wall of the ductus deferens have been sectioned transversely and longitudinally. The single, elongated nuclei are centrally located in the fusiform fiber. The cytoplasm of the cell is of considerable length when seen in longitudinal section. Mitochondria appear as moderately dark rods superimposed on the lighter, cytoplasmic background. 960 ×

2. In longitudinal section, contractile filaments with irregularly spaced mitochondria and smaller, dense bodies occupy most of the cytoplasm. At this magnification clearer regions containing ribosomes and organelles can be identified at the periphery and in a juxtanuclear cone-shaped zone of the smooth muscle cell. 7500 ×

3. L. In arterioles from the uterus the single layer of smooth muscle cells has been sectioned transversely and longitudinally. The muscular layer is separated from the endothelium by a thin, interrupted, dense line, the internal elastic lamina (arrow). 680 ×

4. In transverse section the periphery of the arteriolar smooth muscle fiber is often deeply indented. Dense bodies of homogeneous material extend to a considerable depth into the

cytoplasm from these indentations. Dense bodies of the same nature are randomly distributed among the myofilaments, which appear as a multitude of very fine points or short lines. Most of the organelles and ribosomes are located in either the peripheral or juxtanuclear zones. A basal lamina is located at the base of the endothelium (E). A thin circumferential lamina is closely applied to the muscle fiber. 11,500 ×

5. Muscularis—pyloric canal. A number of oval or elongated dense bodies are located among the contractile filaments in the cytoplasm. In contraction the cell borders are markedly irregular. Collagen, embedded in dense material, and elastic fibers occupy the intercellular space. 5600 ×

6. The internal filaments are believed to transmit the force of their contraction to the plasma membrane of the fiber by an oblique attachment to surface. The majority are believed to act through the intermediary of the dense bodies, which have been seen extending into the cell. In this figure a wisp of filaments appears to reach the surface in one fiber, while in the adjacent cell a broad band of filaments, with transversely sectioned dense bodies included, appears in close proximity to the plasma membrane. 45,000 ×

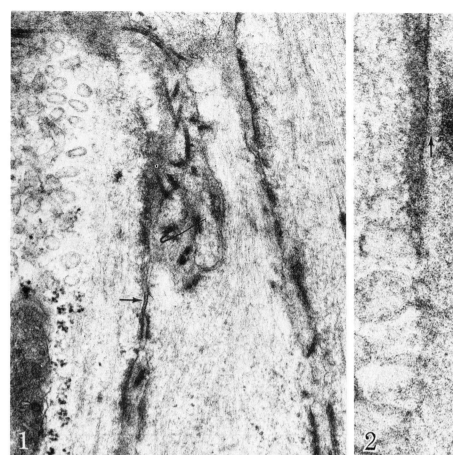

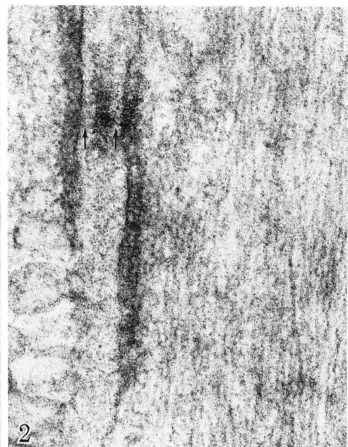

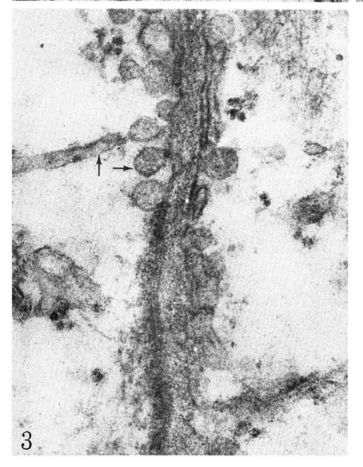

Smooth Muscle—Vas Deferens

1. Short segments of tight junction (arrow) are infrequently seen between smooth muscle cells. These may play an important role in the transmission from cell to cell of the stimulus required for initiation of the contraction. However, these tight junctions are quite uncommon, and transmissions must occur at other types of junctional complexes as well. 160,000 ×

2. Filaments are disposed transversely (arrows) between smooth muscle cells. These may indicate a certain amount of structural continuity between cells at other than the tight junctions. Although the interspace between the muscle cells is considerably greater than in the classical desmosome, the dense cytoplasmic plaques are reminiscent of this feature of the desmosome. The extracellular dense body appears to be an elastic fiber sectioned transversely. 140,000 ×

3. This figure demonstrates fine parallel lines in the surfaces of vesicles and in one of the tubelike processes of the endoplasmic reticulum (arrows). Filaments project from the surface of one of the vesicles. 110,000 ×

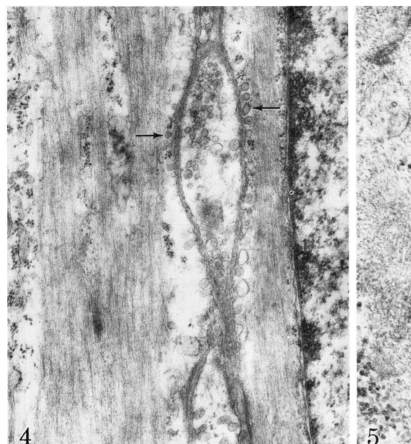

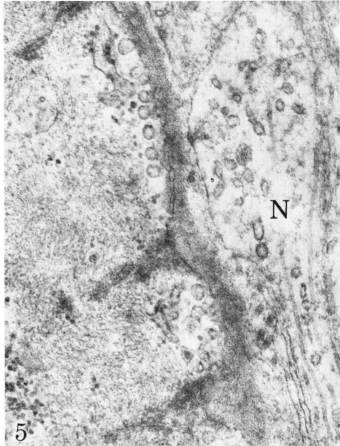

4. Vas deferens. A nerve ending between two muscle fibers contains both dense and clear vesicles. A number of the membrane invaginations in the muscle fiber adjacent to the nerve terminal have a dense content as well (arrows). 32,000 ×

Smooth Muscle Innervation

5. In the wall of an arteriole in the tongue, an autonomic nerve (N) expands on a transversely sectioned smooth muscle fiber. Cytoplasmic microtubules extend into the dilatation of the nerve. The synaptic cleft is approximately 1000 Å in width. The circumferential lamina of homogeneous dense material about the muscle is continuous through the synaptic cleft. Surface invaginations of the smooth muscle are numerous. Enlargements such as this appear along unmyelinated nerve fibers and allow a single nerve to innervate several muscle cells. 60,000 ×

6. Thyroid. Nerve fibers terminate on a smooth muscle fiber in the wall of an arteriole. These endings are partially enveloped in Schwann cell cytoplasm. A few larger vesicles with dense content are included in the nerve ending. A concentration of surface membrane invaginations is a characteristic feature in this area of the membrane of the receptor, the smooth muscle fiber. The synaptic cleft is much wider between autonomic nerve endings and smooth muscle ($\propto 1000$ Å) than between peripheral nerves and skeletal muscle ($\propto 500$ Å). The nerve endings are not invaginated into the surfaces of smooth muscle as is the case with skeletal muscle. 60,000 ×

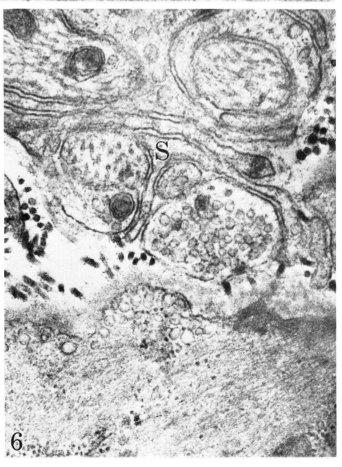

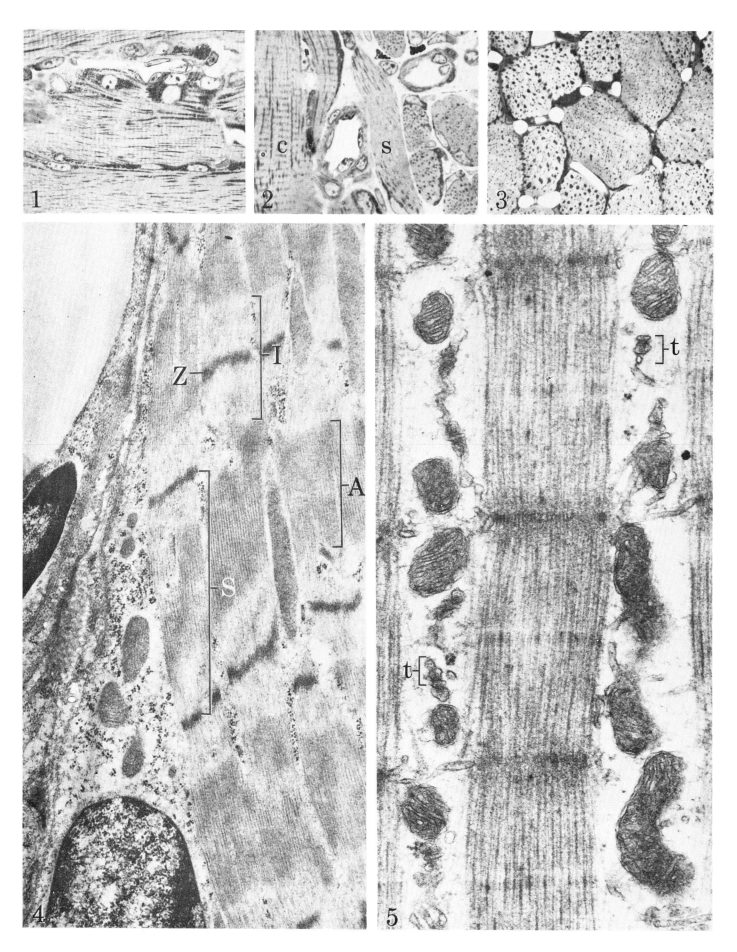

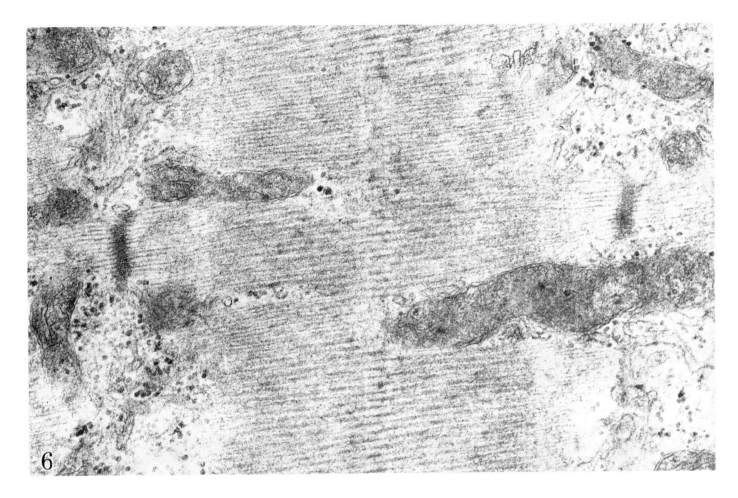

6

Skeletal Striated Muscle

1. L. Tongue. Skeletal muscle fibers are long and multi-nucleated. The nuclei are located at the periphery of the cell and are often embedded in masses of densely stained mitochondria. Other mitochondria are found between individual myofibrils of the cell. The myofibrils are striated in appearance, having alternate pale and dense bands at right angles to their long axes. 450 ×

2. L. In the muscle of the tongue, contracted (c) and stretched (s) fibers, sectioned longitudinally as well as transversely, are found in one field. In stretched striated muscle fibers a lighter I band is found between the darker A bands. In the contracted muscle little is seen of the I band in light microscopy. In the stretched muscle a lightly stained I band exhibits a central dense line. In the transversely sectioned muscle the peripherally located multiple nuclei and the dense mitochondria between the myofibrils can be clearly distinguished. Small arterioles, venules, and capillaries are located between the muscle fibers. 600 ×

3. L. Transversely sectioned striated muscle fibers—diaphragm. The network of lines and the more dense bodies are mitochondria (sarcosomes). The abundant blood supply of skeletal muscle is indicated by the number of capillaries between the transversely sectioned fibers. 450 ×

4. Parts of a muscle fiber and an endothelial cell are separated by a narrow space. The sarcoplasm surrounding the nucleus contains mitochondria and a number of ribosomes. In this stretched skeletal muscle fiber, the darker A and the lighter I bands of the myofibrils are sharply demarcated. The dense transverse lines in the center of the I bands are termed Z lines. The segment between two Z lines is a sarcomere (S). Mitochondria and ribosomes are located between individual myofibrils. 35,000 ×

5. In a contracted myofibril, the I band has been incorporated into the adjacent A bands. This has been accomplished by a sliding of the thinner filaments in the I band between the thicker filaments of the A band. The sarcomere, the segment between two Z lines, has been shortened to little more than the length of an A band. The agranular endoplasmic reticulum is tubular. Two of these tubules abut against a central tubule, which is an invagination of the plasma membrane. These groups of three, seen in many places between the myofibrils, have been termed triads (t). 70,000 ×

Stretched Skeletal Muscle

6. The A band is made up of myosin filaments, termed the primary filaments, approximately 150 Å in diameter. A lighter zone located in the center of the A band has been termed an H zone. There is a darker segment in each primary filament in the center of the H zone. Together, these darker segments across the width of the myofibril have been termed the M line. In both extremities of the A band there is an overlapping of primary filaments with secondary filaments from the I band. 70,000 ×

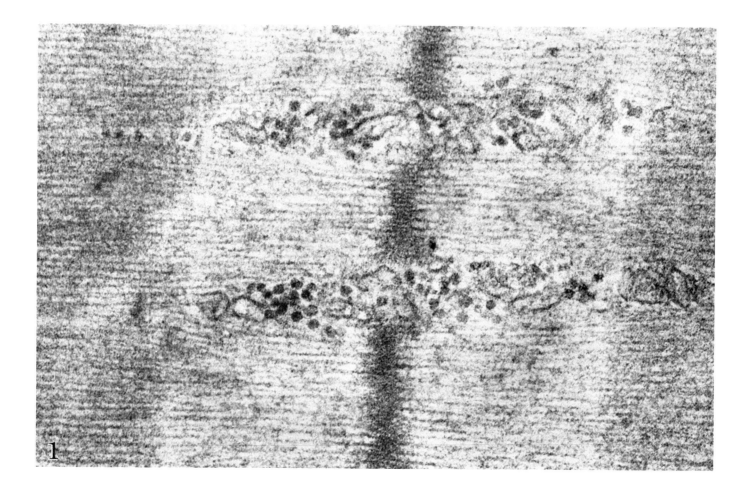

Stretched Skeletal Muscle

1. In a stretched muscle fiber, it often appears that, within the Z line, there is an overlapping of the secondary filaments from the two halves of the I band. The secondary filaments extend for only a short distance between the primary filaments in the A band. Ribosomes and elements of the sarcoplasmic reticulum are found between the myofibrils. 155,000 ×

2. A transverse section of a stretched muscle where the bands of neighboring myofibrils were not in perfect register. A section through primary filaments of the A band, through primary and secondary filaments at the border of the A band and secondary filaments in a section through the I band have all been obtained in one field. Six secondary filaments form a hexagon (h) around each primary filament. Two secondary filaments are located between each pair of adjacent primary filaments. The primary filaments are arranged in triangular array. In some there is a central light zone surrounded by a darker periphery (arrows). Lightly stained fibrillar radiations extend out from many of the primary filaments. Mitochondria, often termed sarcosomes, with densely stained matrices and tubes of agranular endoplasmic reticulum, the sarcoplasmic reticulum, are interposed between the myofibrils. One tube of the sarcoplasmic reticulum contains a considerable quantity of moderately densely stained material. 95,000 ×

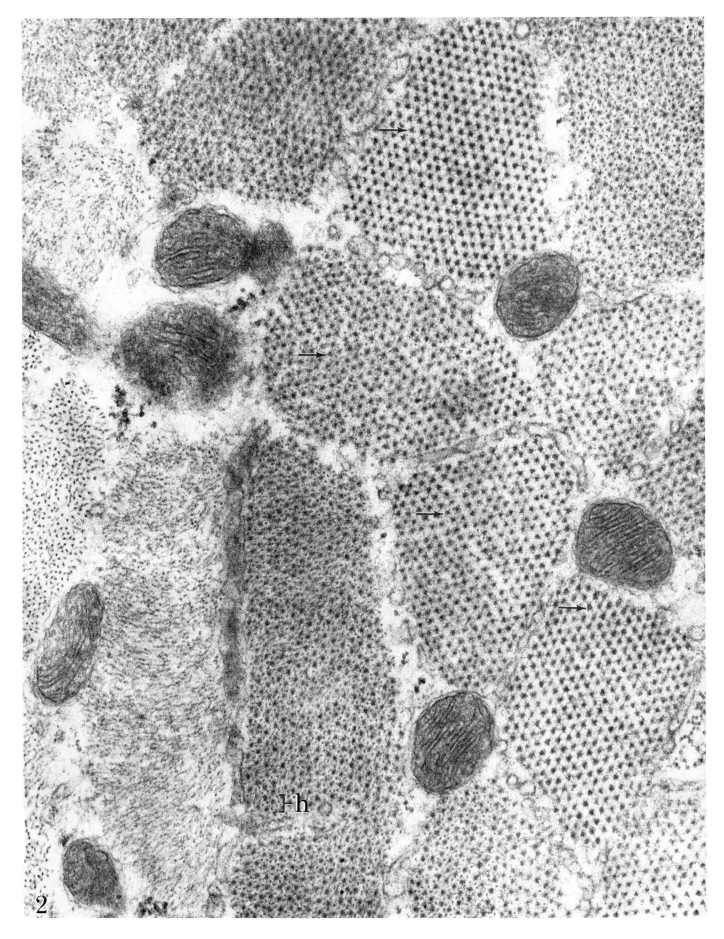

2

⊢h

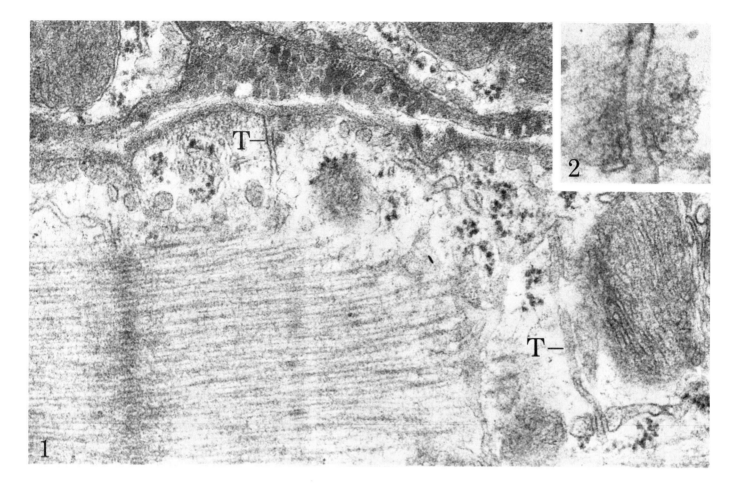

Skeletal Muscle

1. The membrane of the central tubular element of the triad, the T system (T), is continuous with the surface membrane of the muscle fiber. Long T tubules penetrate to triads at a much greater depth. Before the continuity of the T system with the plasma membrane was demonstrated it had been found that inert labels could be induced to enter these tubules, thus indicating that there was a functional continuity with the surface. The rarity with which one finds the tubule opening onto the surface suggests that these may not be open at all times. It is quite possible that permanent structural continuity exists, but the openings onto the surface are only temporary. 100,000 ×

2. Longitudinally and transversely sectioned filaments can be identified in the interspace between the central T tubule and the lateral sarcotubular elements of the triad. 18,000 ×

The Neuromuscular Junction—Diaphragm

3. L. Motor end plates can be demonstrated on skeletal muscles by histochemical localization of cholinesterases in the neuromuscular junction. In this instance the histochemical reaction was followed by a Bodian stain for nerve fibers with the resulting demonstration of the single nerve fiber leading to each end plate. (Contributed by Dr. Pierre Bois.) 375 ×

4. L. A motor end plate is present on each muscle fiber. These end plates are located in a narrow band at approximately the center of the muscle fibers in the diaphragm. In surface views they appear as ramifying structures with a density at the periphery of the branches. In some instances the end plate is seen in lateral view and appears much more dense due to the superimposition of the various parts upon one another. 150 ×

5. L. In light microscopy the motor end plate on skeletal muscle appears as a series of convoluted processes. The processes have a pale center and a dark periphery. The end plate is applied to the surface of muscle fibers. 600 ×

6. In electron microscopy it has been shown that axon terminals, which make up the neural part of the end plate, are embedded in the surface of the muscle fiber. The plasma membrane of the muscle is separated from that of the nerve process by a primary cleft approximately 500 Å in width. The surface of the muscle is again indented by secondary membrane invaginations which extend for considerable distances into the sarcoplasm. Many of these secondary "clefts" branch within the muscle. The plasma membrane of the muscular component of the neuromuscular junction appears considerably more dense than does that of the neural component. This feature is similar to that found in synapses. This added density on the inner surface of the membrane extends part of the way down into the secondary clefts. A number of mitochondria, ribosomes, and elements of the endoplasmic reticulum are found in the cytoplasm of the muscle. There are many mitochondria and synaptic vesicles in the nerve endings. 20,000 ×

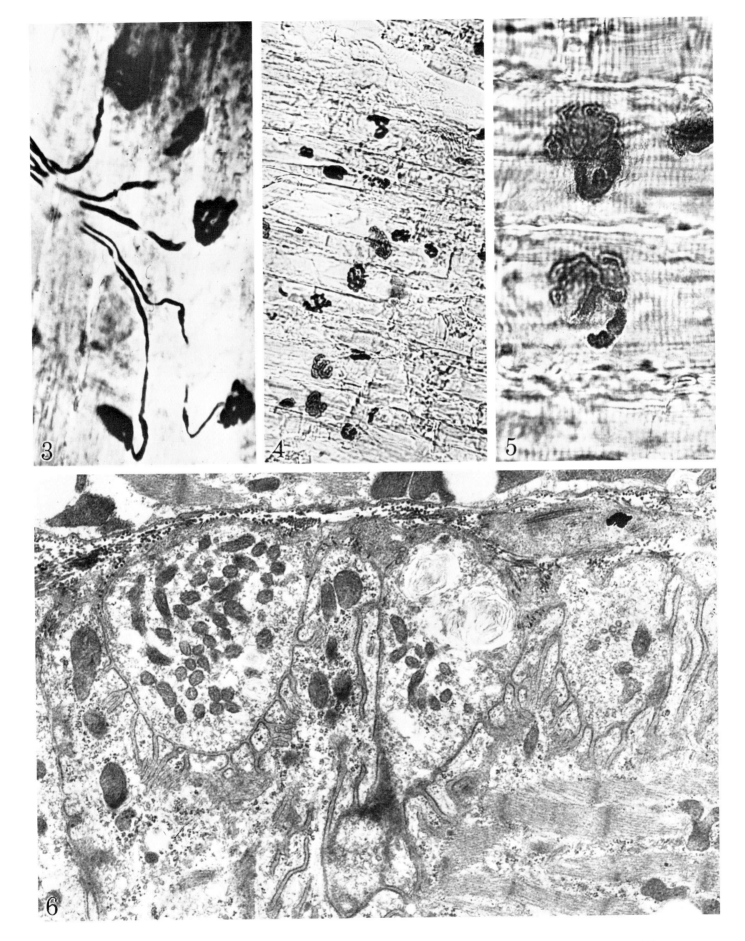

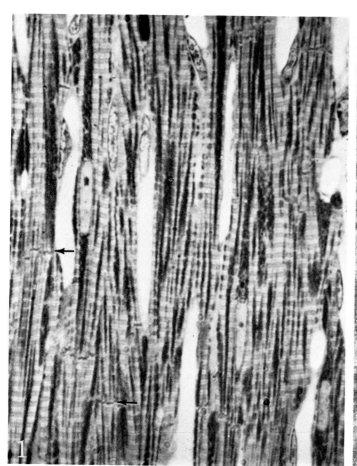

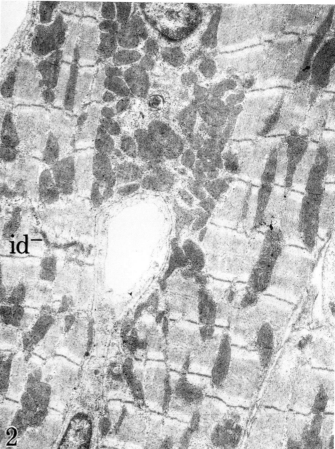

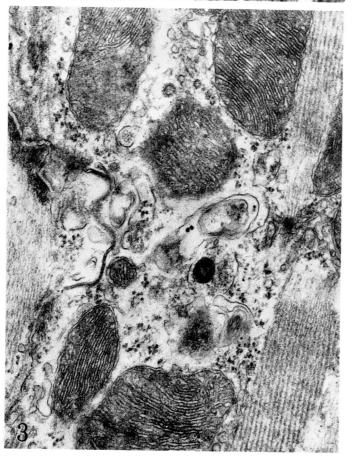

Cardiac Ventricular Muscle

1. L. In cardiac muscle the fibers, with centrally located single nuclei, are arranged in a branching network which was formerly considered a syncytium. The fibers are typically striated with A and I bands, and, even at this level, Z lines and H zones within the fibers can be distinguished. Large numbers of mitochondria extend from the poles of the elongated nuclei. Denser bands (arrows) traverse what would otherwise appear to be continuous muscle fibers. These bands, termed intercalated discs, are intercellular junctional complexes between the ends of fibers. Numerous capillaries from which the blood has been removed by perfusion are found between the groups of fibers. 800 ×

2. In electron microscopy the myofibrils of stretched cardiac muscle fibers have a banded appearance similar to skeletal muscle. Between the myofibrils and in· the juxtanuclear zone are many mitochondria (sarcosomes). The intercalated disc (id), which extends irregularly across the muscle mass, is an intercellular junctional complex between the ends of adjacent muscle fibers, one of which has bifurcated. The thick type of endothelium lines the capillary, which is partially surrounded by a single muscle fiber. 6250 ×

3. On occasion one finds a tight junction as well as desmosomal-type intercellular junctional complexes in an intercalated disc. It is believed that a transmission of impulses between cells may be facilitated in these areas of tight junction. A coated vesicle with a dense peripheral content and a dense granule are found in the cytoplasm. The dense granules are uncommon in ventricular muscle. 50,000 ×

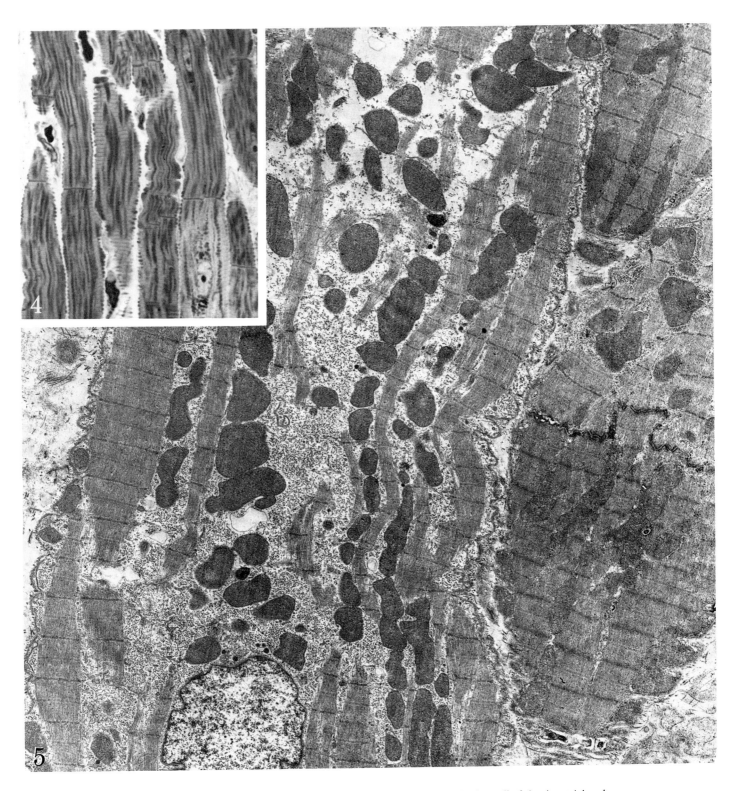

Conducting Fibers, Sinuatrial Node—Dog Heart

4. L. In the sinuatrial node, variations in content of myofibrils and in the surfaces of the fibers are evident. An indented surface is seen on many of these fibers. The nuclei are centrally placed and are quite irregular in outline. The myofibrils in some cells appear widely separated and, in these intervals, moderately stained mitochondria are located. Intercalated discs are prominent between the ends of the individual muscle fibers. 600 ×

5. In a specialized conducting cell of the sinuatrial node a considerable amount of cytoplasm with numerous ribosomes, a number of mitochondria, and dense granules are found between the myofibrils. The myofibrils are in a state of contraction. Indentations in the sarcolemma of the conducting fiber occur opposite the Z lines. Mitochondria are contained in some of the protrusions between the surface indentations. The comparatively smooth surface and the relatively large number of myofibrils in ordinary contractile fibers is obvious in the adjacent cells. Two of these are interconnected by an extensive intercalated disc. 7000 ×

REFERENCES

Andersson-Cedergren, E. (1959). Ultrastructure of motor end plate and sarcoplasmic components of mouse skeletal muscle fiber as revealed by three-dimensional reconstructions from serial sections. *J. Ultrastruct. Res.* Suppl. **1**, 1–191.

Barnett, R. J. (1962). The fine structural localization of acetyl-cholinesterase at the myoneural junction. *J. Cell. Biol.* **12**, 247–262.

Coërs, C. (1967). Structure and organization of the myoneural junction. *Intern. Rev. Cytol.* **22**, 239–267.

Hess, A. (1965). The sarcoplasmic reticulum, the T system, and the motor terminals of slow and twitch muscle fibers in the garter snake. *J. Cell. Biol.* **26**, 467–476.

Hirano, H., and Ogawa, K. (1967). Ultrastructural localization of cholinesterase activity in nerve endings in the guinea pig heart. *J. Electronmicroscopy* (*Tokyo*) **16**, 413–421.

Huxley, H. E. (1960). Muscle cells. *In* "The Cell" (J. Brachet, and A. E. Mirsky, eds.), Vol. 4, pp. 365–481. Academic Press, New York.

Huxley, H. E. (1964). Evidence for continuity between the central elements of the triads and extracellular space in frog sartorius muscle. *Nature* **202**, 1067–1071.

Jamieson, J. D., and Palade, G. E. (1964). Specific granules in atrial muscle cells. *J. Cell Biol.* **23**, 151–172.

Kelly, R. E. (1968). Localization of myosin filaments in smooth muscle. *J. Cell Biol.* **37**, 105–116.

Lane, B. P. (1965). Alterations in cytological detail of intestinal smooth muscle cells in various stages of contraction. *J. Cell Biol.* **27**, 199–213.

Lane, B. P., and Rhodin, J. A. G. (1964). Cellular interrelationships and electrical activity in two types of smooth muscle. *J. Ultrastruct. Res.* **10**, 470–488.

Merrillees, N. C. R. (1968). The nervous environment of individual smooth muscle cells of the guinea pig vas deferens. *J. Cell Biol.* **37**, 794–817.

Rosenbluth, J. (1965c). Smooth muscle: An ultrastructural basis for the dynamics of its contraction. *Science* **148**, 1337–1339.

Sandborn, E. B., Côté, M. G., Roberge, J., and Bois, P. (1967a). Microtubules et filaments cytoplasmiques dans le muscle de mammifères, *J. Microscopie* **6**, 169–178.

Simpson, F. O., and Devine, C. E. (1966). The fine structure of autonomic neuromuscular contacts in arterioles of sheep renal cortex. *J. Anat.* **100**, 127–137.

Smith, D. S. (1964). Smooth, cardiac, and skeletal muscle. *In* "Electron Microscopic Anatomy" (S. M. Kurtz, ed.), pp. 267–293. Academic Press, New York.

Stanley, H. (1960). The structure of striated muscle as seen by the electron microscope. *In* "Muscle" (G. H. Bourne, ed.), Vol. 1: Structure, Academic Press, New York.

Walker, S. M., and Schrodt, G. R. (1965). Continuity of the T system with the sarcolemma in rat skeletal muscle fibers. *J. Cell Biol.* **27**, 671–677.

Zacks, S. I. (1964). "The Motor Endplate." Saunders, Philadelphia, Pennsylvania.

THE
LYMPHOID ORGANS

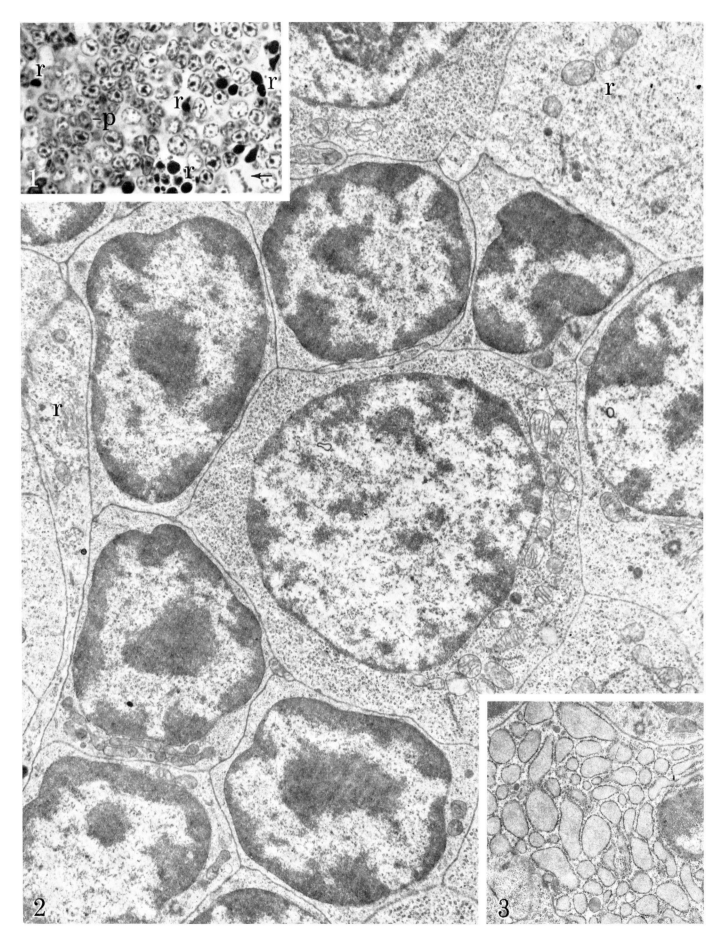

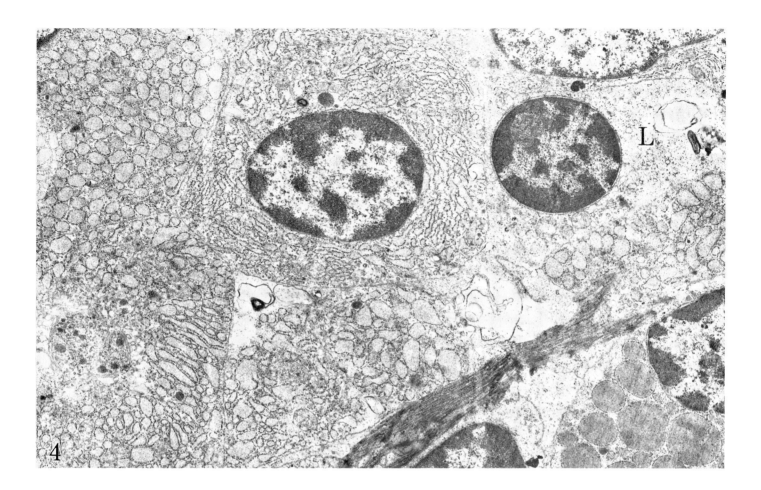

The Thymus

The role of the thymus in the production of lymphocytes and its significance in immunological processes has been recognized only in recent years.

1. L. In the thymus reticular cells (r) with lightly stained nuclei form a supportive cytoplasmic network or reticulum for collections of the much smaller lymphocytes. These large cells contain varying numbers of cytoplasmic dense granules and vesicles. The lymphocytes have more densely stained nuclei and scanty cytoplasm. Plasma cells (p) with more abundant, dense cytoplasm are only occasionally seen in the thymus of the young rat. At the corticomedullary junction reticular cells with large, dense cytoplasmic granules are more frequently encountered than elsewhere in the thymus. The rate of cell division in the thymus is high. One cell in this field is in metaphase (arrow). 500 ×

2. A group of large to small lymphocytes demonstrates the increased concentration of chromatin which accompanies the reduction in size of the nuclei. Most of the cytoplasmic organelles are concentrated at one pole of the lymphocyte, the remainder of the cytoplasm being occupied by large numbers of ribosomes. A small number of dense granules can be seen among the organelles, most commonly at the border of the Golgi complex. The lymphocytes are closely applied to one another except where thin processes of reticular cell cytoplasm (r) intervene.

Moderately dense granules and tubules are common in the reticular cells 13,000 ×

3. Few plasma cells are found in the thymus of very young rats, but their numbers increase with age. In the plasma cell the protein-filled, distended cisternae of the granular endoplasmic reticulum occupies most of the abundant cytoplasm. Some small, moderately dense granules with a smooth surrounding membrane are seen in these cells. 11,000 ×

The Thymus—Aged Rat (24 Months)

4. In the aged rat very few small lymphocytes are present in the thymus. The lymphoid cells appear large, and many plasma cells can be identified. Very large, dense granules and vacuoles in the cytoplasm of phagocytic reticular cells are common findings. In the small cells, which probably represent lymphocytes (L), one finds more granular endoplasmic reticulum and fewer free ribosomes than is the case in the thymic lymphocytes of younger animals. The amount of distension of the cisternae of the granular endoplasmic reticulum in the plasmocytes and the density of their content varies markedly. Dense granules, both small and large, appear in some of the plasma cells. In a part of one cell the cisternae of the granular reticulum appear as rounded vesicles with a moderately dense content. 7000 ×

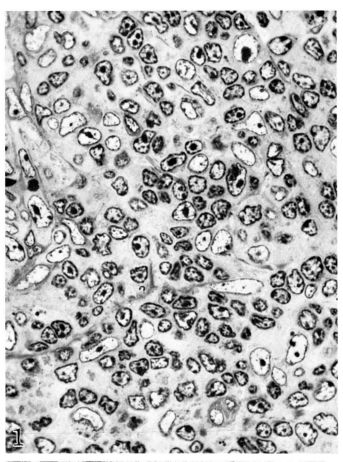

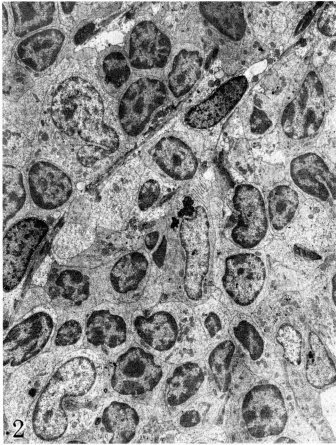

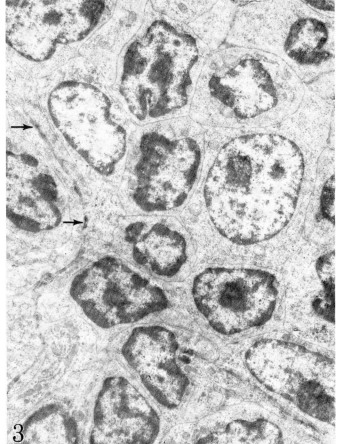

The Spleen

Two distinct types of tissue are present in the spleen, the red pulp and the white pulp. Destruction of aged blood cells, particularly the erythrocytes, occur in the red pulp. Thus this portion of the spleen is distinguished by sinusoids filled with deformed red cells and macrophages which contain numerous hemoglobin filled granules. The spleen of the adult rat is a blood-forming organ. Thus the red pulp contains megakaryocytes, granulocytes, and large numbers of more densely stained cells of the erythropoietic series. The white pulp is made up of diffuse or nodular formations of cells in the lymphoid series.

1. L. The lymphoid portion of the spleen is made up of a mass of large reticular cells, lymphoblasts, and large-to-small lymphocytes. The reticular cells, some of which contain a number of dense granules, are phagocytic cells which also form a supporting network for the collections of lymphocytes. Other immature blood cells can also be seen. 720×

2. A wide range of immature cells in the hemopoietic series, as well as macrophages, with large, dense cytoplasmic granules are seen in the red pulp of the spleen in the rat. 2250×

3. Large, medium, and small lymphocytes are in close contact with one another and with long, thin processes of reticular cell cytoplasm in the white pulp. Occasionally, a wider intercellular space with some collagen fibrils is seen between two of the reticular cell cytoplasmic processes (arrows). 4700×

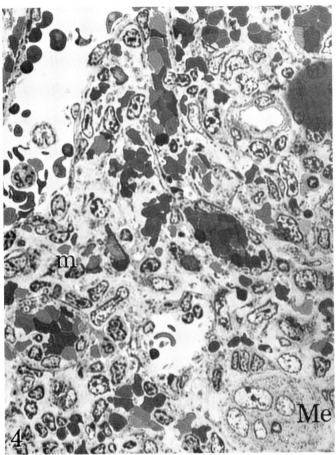

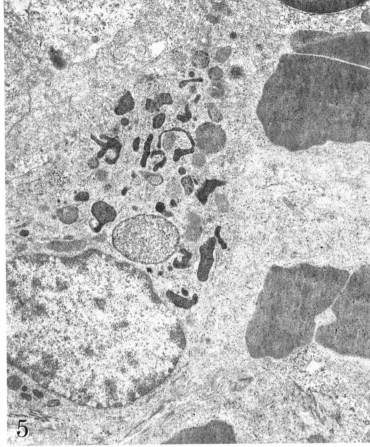

The Spleen—Red Pulp

4. L. Venous sinuses (sinusoids), some packed with deformed erythrocytes, form a branching network through the red pulp. The endothelial lining of the sinusoids is quite thick. In the intervening splenic cords one finds macrophages (m), a megakaryocyte (Me), and a wide range of other myeloid cells. 720 ×

5. Very irregularly shaped, extremely dense granules and others with only a dense periphery are often found among the paler, homogeneous, hemoglobin-filled granules in the macrophages of the splenic cords. 27,500 ×

6. At higher magnification a particulate material is seen in the membrane-bound granules. These fine particles are ferritin. Some granules contain a mixture of hemoglobin with the ferritin particles mostly at the periphery. One can distinguish two membranes with an intervening space surrounding each of these ferritin-containing granules. 75,000 ×

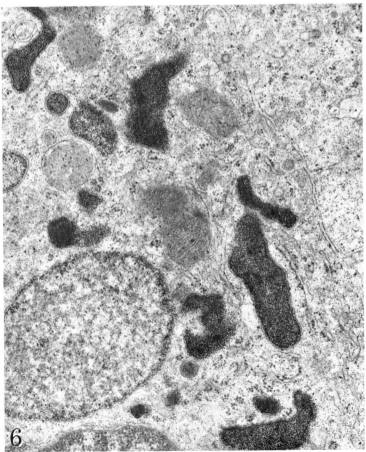

REFERENCES

Bessis, M. (1964). Lymphoid tissue. *In* "Electron Microscopic Anatomy" (S. M. Kurtz, ed.), pp. 183–198. Academic Press, New York.

Bois, P., and Sandborn, E. B. (1967). Morphologie d'un lymphosarcome thymique expérimental chez le rat. *Compt. Rend. Assoc. Anat.* **137**, 245.

Bois, P., Sandborn, E. B., and Messier, P. E. (1969). A study of thymic lymphosarcoma developing in magnesium-deficient rats. *Cancer Res.* **29**, 763.

Burnet, S. M. (1962). The thymus gland. *Sci. Am.* **207**, 50–57.

Clark, S. L. (1966). Cytological evidences of secretion in the thymus. *Thymus: Exptl. Clin. Studies Ciba Found. Symp.,* pp. 32–38.

Defendi, V., and Metcalf, D. (1964). The thymus. *Thymus Symp., Philadelphia, 1964,* p. 145. Wistar Inst. Press, Philadelphia, Pennsylvania.

Good, R. A., and Gabrielsen, A. E. (1964). "The Thymus in Immunobiology." Harper (Hoeber), New York.

Holyoke, E. A., Latta, J. S., and Volenec McLean, J. (1966). A study of the ultrastructure of the developing spleen in the albino rat. *J. Ultrastruct. Res.* **15**, 87–99.

Kohnen, P., and Weiss, L. (1964). An electron microscopic study of thymic corpuscles in the guinea pig and the mouse. *Anat. Record* **148**, 29–58.

Lundin, P. M., and Schelin, U. (1965). Ultrastructure of the rat thymus. *Acta Pathol. Microbiol. Scand.* **65**, 379–394.

Marchesi, V. T., and Gowans, J. L. (1963). The migration of lymphocytes through the endothelium of venules in lymph nodes: An electron microscope study. *Proc. Roy. Soc.* **B159**, 283–290.

Metcalf, D., and Brumby, M. (1966). The role of the thymus in the ontogeny of the immune system. *J. Cellular Physiol.* **67**, Suppl. 1, 149–168.

Murray, R. G., Murray, A. S., and Pizzo, A. (1965). The fine structure of mitosis in rat thymic lymphocytes. *J. Cell Biol.* **26**, 601–619.

Nossal, G. J. V. (1964). How cells make antibodies. *Sci. Am.* **211**, 106–115.

Oehmke, H.-J. (1968). Periphere Lymphgefässe des Menschen und ihre funktionelle Struktur. *Z. Zellforsch. Mikroskop. Anat.* **90**, 320–322.

Orlic, D., Gordon, A. S., and Rhodin, J. A. G. (1965). An ultrastructural study of erythropoietin-induced red cell formation in mouse spleen. *J. Ultrastruct. Res.* **13**, 516–542.

Sainte-Marie, G., and Sin, Y. M. (1968). Structures of the lymph node and their possible function during the immune response. *Rev. Can. Biol.* **27**, 191–207.

Tooze, J., and Davies, H. G. (1967). Light- and electron-microscope studies on the spleen of the newt triturus Cristatus: The fine structure of erythropoietic cells. *J. Cell Sci.* **2**, 617–640.

Van Haelst, U. (1967). Light and electron microscopic study of the normal and pathological thymus of the rat. I. The normal thymus. *Z. Zellforsch. Mikroskop. Anat.* **77**, 534–553.

Weiss, L. (1963). Electron microscopic observations on the vascular barrier in the cortex of the thymus of the mouse. *Anat. Record* **145**, 413–437.

Welsh, R. A. (1966). Kurloff body formation in the guinea pig lymphocyte. *J. Ultrastruct. Res.* **14**, 556–570.

Yamada, E. (1957). The fine structure of the megakaryocyte in the mouse spleen. *Acta. Anat.* **29**, 267–290.

Yoffey, J. M. (1964). The lymphocyte. *Ann. Rev. Med.* **15**, 125–148.

THE CIRCULATORY
SYSTEM

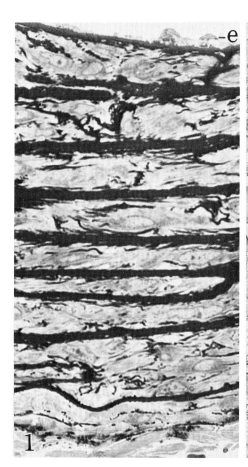

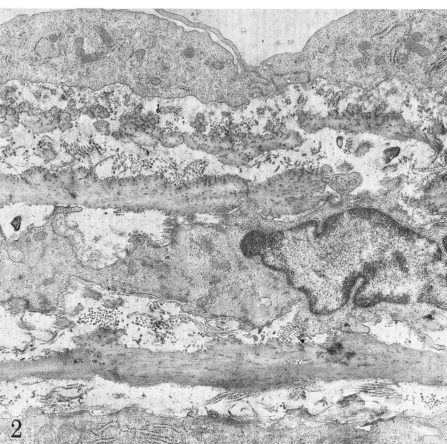

The Circulatory System

The features of the endothelial cell which lines all blood vessels have been considered in the chapter on fundamental tissues in chapter 2. Cardiac muscle and smooth muscle, the contractile elements of the circulatory system, have been considered in the chapter on muscle. This chapter includes special features of large arteries, arterioles, venules, capillaries, and small lymphatics.

1. L. Rat aorta. The media of this very large vessel is made up of many layers of smooth muscle separated by elastic laminae. In young adult animals these laminae appear interrupted in some locations. The endothelium (e) appears as a very thin layer lining the lumen. 640 ×

2. Rabbit aorta. In the electron microscope the endothelium can be seen to be moderately thick with a considerable amount of granular endoplasmic reticulum and numerous plasma membrane invaginations. A long cytoplasmic operculum projects into the lumen from a point adjacent to the intercellular junction. The internal elastic lamina is interrupted, being made up of numerous elastic fibers about which both filaments and more distinct collagen fibrils can be identified. A smooth muscle cell contains a considerable amount of granular endoplasmic reticulum. 14,000 ×

Arteriole

3. Kidney. In a contracted arteriole the endothelium becomes heaped up and the nuclei appear to protrude into the lumen. An almost continuous elastic lamina separates the basal lamina of the endothelial cells from the circumferential lamina of the smooth muscle cells. A number of granules in addition to the dense bodies can be seen in the smooth muscle fibers. An interrupted layer of elastic fibers is seen at the outer limit of the muscularis. This could be considered a discontinuous external elastic lamina. 11,000 ×

4. Pancreas. In the contracted, very small arteriole the endothelium becomes quite folded. Whether this change in height is due purely to the contraction of the surrounding smooth muscle cells (sm) or partially to contractile forces within the endothelial cell itself is difficult to determine. In these small arterioles the elastic laminae are no longer present. 18,000 ×

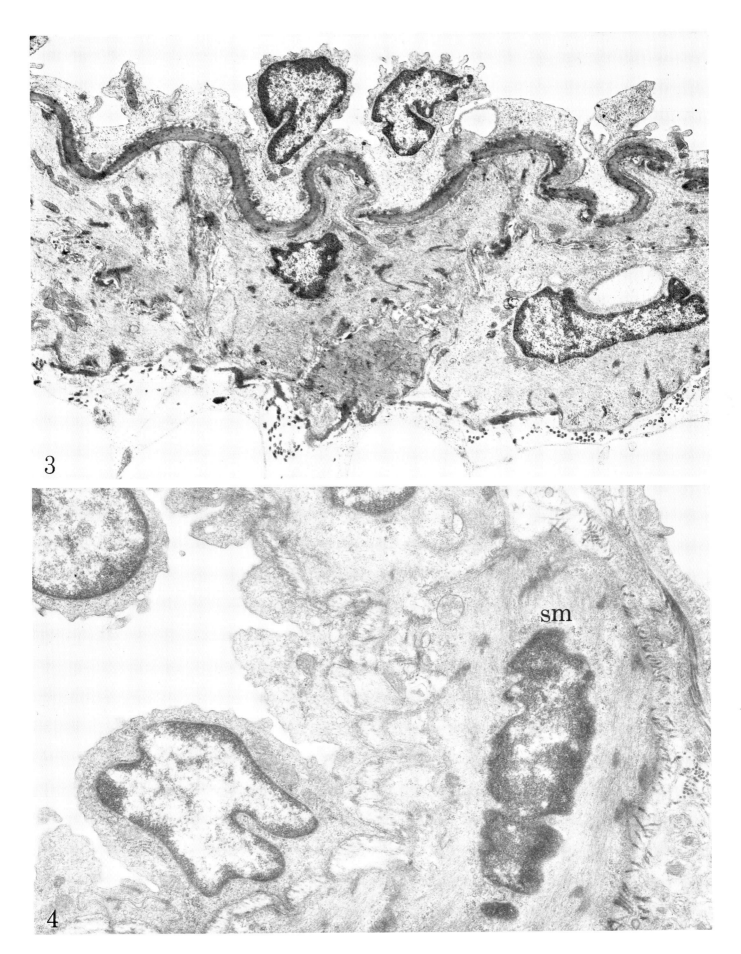

3

4

sm

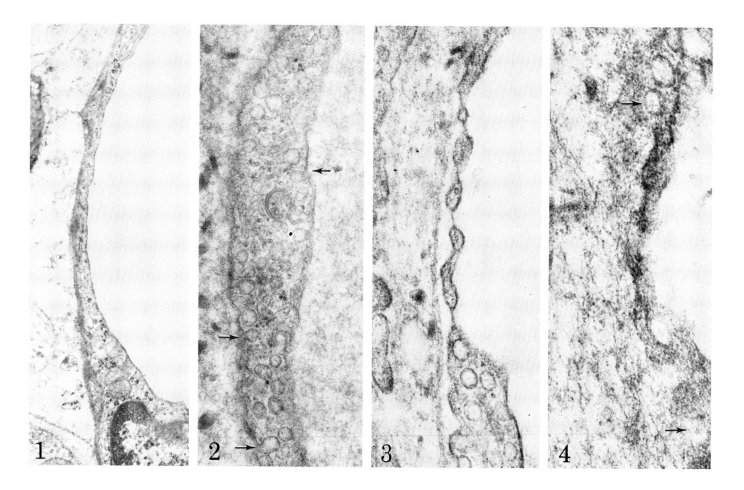

Capillary

1. Capillary—cerebellum. A single flattened endothelial cell is able to completely encircle the lumen of a capillary. In the central nervous system, with the exception of neurosecretary appendages, the endothelial cell cytoplasm is rarely of the very thin, fenestrated type. The elongated nucleus occupies a thickened part of the capillary wall. The cell rests on a well-defined basal lamina. 16,500×

2. Mammary gland. Endothelial cells, active in the transport of fluids and metabolites, demonstrate a very large number of vesicular-appearing profiles in the cytoplasm. Deep invaginations of the basal plasma membrane (arrows), as well as a number of filaments and cytoplasmic microtubules, are common features of these cells. 62,000 ×

3. The fenestrated cytoplasm is actually a very thin sheet with circular pores. In section the cytoplasm surrounding the pores appears as a chain of globules interconnected by the thin lines or diaphragms which span the pores. 75,000 ×

4. When sectioned obliquely or tangentially, the capillary pores appear as circular openings in the cytoplasm with a central density (arrows). 62,000 ×

Lymphatics and Postcapillary Venules of the Ileum

5. L. Many lymphocyte-filled lymphatics (l), capillaries (c), and postcapillary venules (v) are found between the crypts and the border of the Peyer's patch (P) of lymphoid tissue located in the wall of the ileum. The endothelium of the postcapillary venules in lymphoid tissues of normal animals often appears hyper-

trophied. In these locations diapedesis of small lymphocytes across the endothelium is more evident than elsewhere. When the venule is located between the base of the crypt and the Peyer's patch of the ileum, the hypertrophy takes place in the portion of the wall which faces the lymphoid tissue. A normal, flattened endothelium remains on the mucosal aspect. This thickening of the postcapillary venule endothelium, in association with this small lymphocyte migration, is accentuated after the administration of antigens to animals. It is not known whether this response is a direct result of antigen stimulation or an indirect response elicited by the small lymphocytes. 130 ×

6. L. At a higher magnification the thickening of the tunica intima of the postcapillary venule (v) can be seen to be partly due to hypertrophy of the endothelial cells and partly to an invasion of lymphocytes between the basal lamina and the endothelial cells. The lymphatic vessels (l) have a flattened endothelium. The tunica media of both the postcapillary venule and the lymphatic are made up of a very thin layer of smooth muscle. 640 ×

7. The lymph vessel (l), with many lymphocytes in its lumen, has a thin endothelial layer. Smooth muscle cells are widely separated in the media of the lymph vessel, as indicated by the single fiber. The endothelium of the postcapillary venule (v) increases abruptly in thickness in the wall facing the lymphoid tissue. The endothelial cells of the venule contain irregular dense granules. Lymphocytes have invaded the intima and are located between the endothelium and its basal lamina. A single interrupted layer of smooth muscle surrounds the venule. 7000 ×

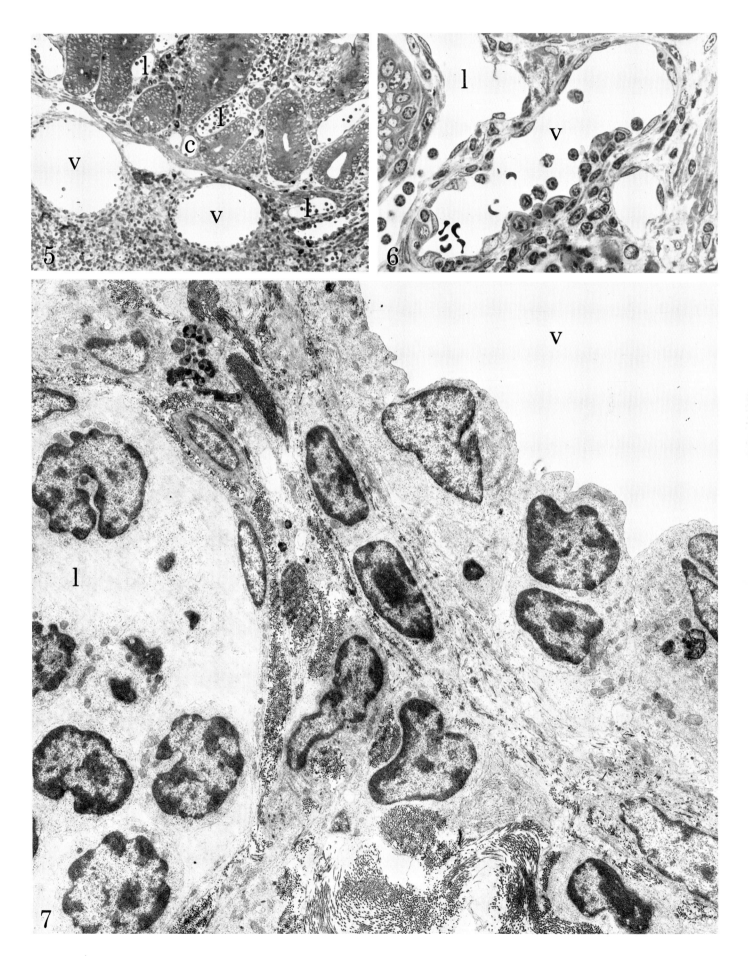

REFERENCES

Bartman, J. (1968). Ultrastructure of elastic tissue of the newborn rat aortic media. *J. Microscopie* 7, 355–365.

Bennett, H. S., Luft, J. H., and Hampton, J. C. (1959). Morphological classifications of vertebrate blood capillaries. *Am. J. Physiol.* 196, 381–390.

Bruns, R. R., and Palade, G. E. (1968). Studies on blood capillaries. 1. General organization of blood capillaries in muscle. *J. Cell Biol.* 37, 244–276.

Buck, R. C. (1958). The fine structure of endothelium of large arteries. *J. Biophys. Biochem. Cytol.* 4, 187–190.

Buck, R. C. (1961). Intimal thickening after ligature of arteries. An electron microscopic study. *Circulation Res.* 9, 418–426.

Fawcett, D. W. (1963). Comparative observations on the fine structure of blood capillaries. *In* "Peripheral Blood Vessels" (J. L. Orbison and D. E. Smith, eds.), *Intern. Acad. Pathol. Monograph* No. 4, pp. 17–44. Williams & Wilkins, Baltimore, Maryland.

Florey, L. (1966). The endothelial cell. *Brit. Med. J.* II, 487–490.

Friederich, H. H. (1968). The tridimensional ultrastructure of fenestrated capillaries. *J. Ultrastruct. Res.* 23, 444–456.

Fuchs, A., and Weibel, E. R. (1966). Morphometrische untersuchung der verteilung einer spezifischen cytoplasmatischen organelle in endothelzellen der rate. *Z. Zellforsch. Mikrskop. Anat.* 73, 1–9.

Hoff, H. F. (1966). An electron microscopy study of phosphatase activity in the endothelial cells of rabbit aorta. *J. Histochem. Cytochem.* 14, 719–724.

Karnovsky, M. J. (1967). The ultrastructural basis of capillary permeability studies with peroxidase as a tracer. *J. Cell Biol.* 35, 213–236.

Landis, E. M., and Pappenheimer, J. R. (1963). Exchange of substances through the capillary walls. *In* "Handbook of Physiology" (Am. Physiol. Soc., J. Field, ed.), Sect. 2, Vol. II, p. 961–1034. Williams & Wilkins, Baltimore, Maryland.

Luft, J. H. (1966). Fine structure of capillary and endocapillary layer as revealed by ruthenium red. *Federation Proc.* 25, 1773–1783.

Majno, G. (1965). Ultrastructure of the vascular membrane. *In* "Handbook of Physiology" (Am. Physiol. Soc., J. Field, ed.), Sect. 2, Vol. III, p. 2293–2375. Williams & Wilkins, Baltimore, Maryland.

Marchesi, V. T., and Barrnett, R. J. (1963). The demonstration of enzymatic activity in pinocytic vesicles of blood capillaries with the electron microscope. *J. Cell Biol.* 17, 547–556.

Marchesi, V. T., and Barrnett, R. J. (1964). The localization of nucleoside phosphatase activity in different types of small blood vessels. *J. Ultrastruct. Res.* 10, 103–115.

Paule, W. J. (1963). Electron microscopy of the newborn rat aorta. *J. Ultrastruct. Res.* 8, 219–235.

Rhodin, J. A. G. (1967). The ulstrastructure of mammalian arterioles and precapillary sphincters. *J. Ultrastruct. Res.* 18, 181–223.

Rohlich, P., and Olah, I. (1967). Cross-striated fibrils in the endothelium of the rat myometral arterioles. *J. Ultrastruct. Res.* 18, 667–676.

Winckler, G., and Foroglou, C. (1966). Structure fine des cellules endothéliales des artérioles du cerveau chez l'homme. *Z. Anat. Entwicklungsgeschichte* 125, 245–254.

Yamamoto, T. (1967). Circulatory organ. *In* "Fine Structure of Cells and Tissues Electron Microscopic Atlas" (E. Yamada, T. Yamamoto, and Y. Watanabe, eds.), Vol. I, pp. 81–143. Igaku Shoin, Tokyo.

THE DIGESTIVE SYSTEM

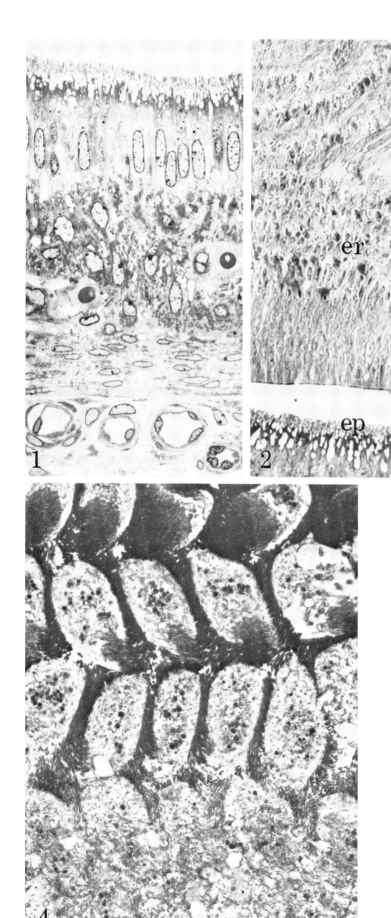

The Tooth—Enamel Organ

Deposition of enamel, the hardest material in the body, follows the deposition of dentine in a basal lamina between the ameloblasts, of epidermal origin, and the odontoblasts, of mesodermal origin. The incisor teeth of the rat continue to grow throughout life.

1. L. In the enamel organ, which has been separated from the lower incisor tooth, one finds: (a) tall columnar cells, the ameloblasts; (b) a narrow compact layer, the stratum intermedium; and (c) a looser arrangement of stellate reticular cells. A closely packed layer of fibroblasts and collagen make up the periodental membrane of the alveolus. The ameloblasts are very tall with moderately large, elongated nuclei. A number of dense granules are located in the juxtanuclear cytoplasm and large vesicles develop in the cell pseudoapex. Thin cytoplasmic processes extend from the cell pseudoapex for a considerable distance toward the developing tooth (which is not included). This pole of the ameloblast, which has the appearance of an apex, is developmentally the base of the cell. 640×

2. L. The processes of the ameloblasts which project beyond the desmosome layer are termed Tomes' enamel processes (ep). The extracellular layer of enamel rods (er), or prisms, has been separated from these processes during tissue preparation. This shows a zone of rods, parallel to one another, in the part adjacent to the Tomes' processes. Rods at right angles to one another are seen at a deeper level. 640×

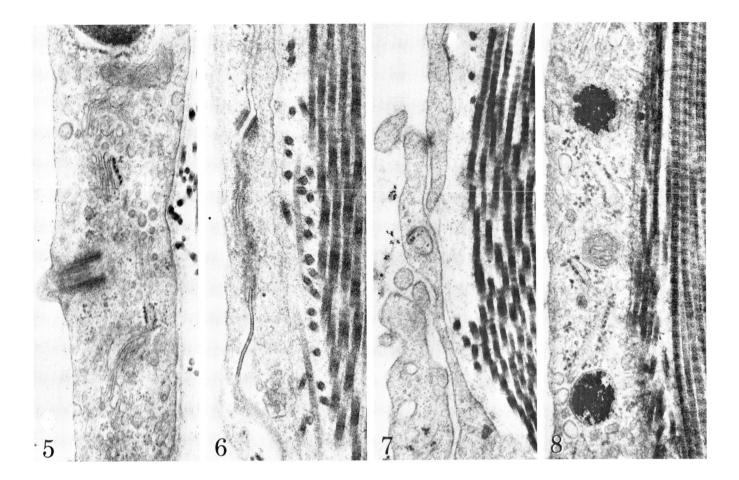

5 6 7 8

The Tooth—Enamel Formation

3. At a later stage of mineralization the density of the apatite can be seen on the fibrils within the rods. The cytoplasmic extensions of the ameloblasts (Tomes' enamel processes) can be seen as distinctly membrane limited. This type of evidence has led to the realization that enamel formation is extracellular. No enamel surrounds the Tomes' processes at this stage of the mineralization. 35,000 ×

4. As mineralization progresses, the interprismatic enamel envelops the Tomes' processes of the pseudoapices of the ameloblasts, as seen here. It is believed that as the enamel increases the prisms encroach upon the Tomes' processes and eventually the ameloblasts recede in height. 11,250 ×

The Peritoneum

A simple squamous epithelial (mesothelial) layer of the peritoneum lines the abdominal cavity and, with its underlying connective tissue, invests the viscera. It is derived from the mesoderm.

5. The mesothelial cells of the peritoneum have a moderate-sized Golgi complex in a juxtanuclear position and a small amount of granular endoplasmic reticulum. A basal body of a cilium is located near the Golgi apparatus. Membrane invaginations into both surfaces of the cell and a thin basal lamina can be seen. 50,000 ×

6. Both tight junctions and typical desmosomes are found between adjacent mesothelial cells. 55,000 ×

7. In other instances only the desmosomal type of junctions appears to interconnect the adjacent cells which form this layer. In this instance the colloidal gold, injected into the peritoneal cavity, can be seen as very dense particles therein. Some of this colloidal gold has been taken up by the peritoneal cell and is located within the cisternae of the smooth-surfaced endoplasmic reticulum. 50,000 ×

8. After a longer period of time large, membrane-bound organelles become packed with the colloidal gold particles, which appear extremely dense. One of these vesicles is at the end of a Golgi zone, while the other is surrounded by endoplasmic reticulum and surface invaginations. 50,000 ×

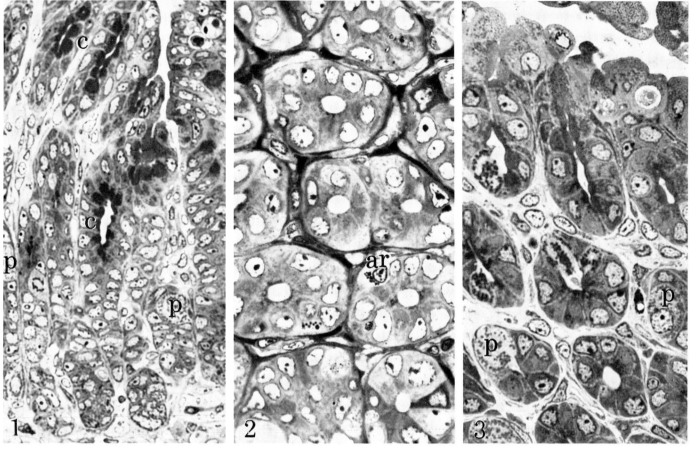

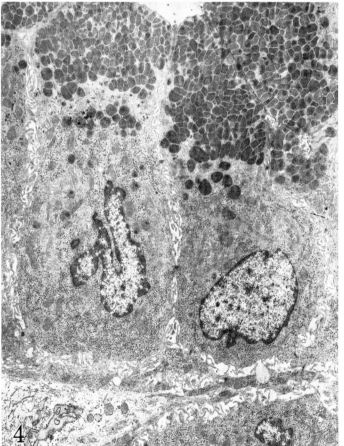

The Stomach

1. L. Long tubular glands empty into the gastric pits. The pits and the necks of the glands are lined by tall columnar mucous cells. Deeper within the glands of the body of the stomach in the rat, these cells are quickly replaced by large parietal (p) or oxyntic cells and chief (c) or zymogen cells. The chief cells, with the dense apical secretory content, are believed to secrete pepsin, rennin, and gastric intrinsic factor. The large parietal cells secrete the hydrochloric acid of the gastric juice. The cell in anaphase is evidence of mitotic activity in the depths of the gland. 375×

2. L. The mucus, stained less intensively than in Fig. 1, appears as an area of light-to-moderate density which occupies most of the supranuclear portion of the mucous neck cells. Cells, with dense granules superimposed on a pale background cytoplasm, and located between the epithelial cells and basal laminae, are in the group generally classified as argentaffin, argyrophil, or enterochromaffin cells. In the stomach most of these have been found to be a distinct cell type, rich in cholinesterase, and have been termed ChE-rich argyrophil cells (ar). 600×

3. L. In the lower part of the neck of the gastric glands, very large parietal cells (p) with discrete, dark mitochondria against a very pale background are found. The basal portion of the parietal, acid-secreting cell is of an exceptional size. 600×

4. The mucous neck cells are simple columnar in shape with irregular nuclei and a markedly convoluted surface. Numerous cytoplasmic processes project from both basal and lateral surfaces. The apices of the cells are filled with irregularly shaped, moderately dense mucus granules. 4500×

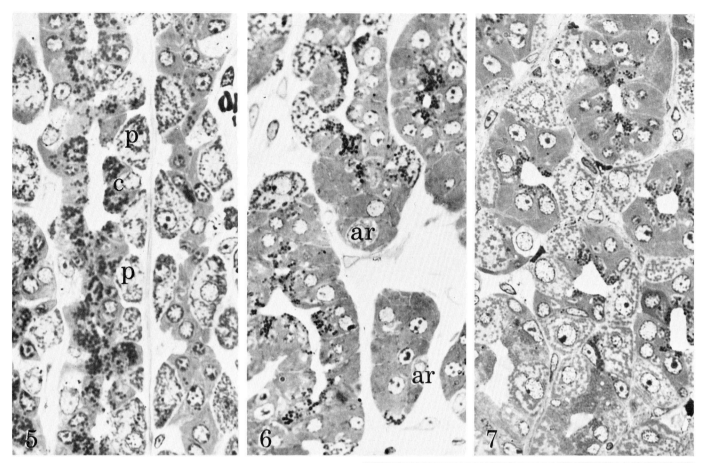

The Stomach Body

5. L. The chief (c) and the parietal cells (p) of the gastric glands are wedge-shaped. The apex of the chief cell is extremely wide. The base of the parietal cell more than compensates for the relative deficiency in the size of the base of the chief cell. The narrow neck of the parietal cell reaches the lumen of the gland in a depression between adjacent chief cells. There has been some difficulty in demonstrating these features in thick sections. Densities throughout the parietal cell are very large mitochondria, while the more distinct circular, apical densities in the chief cell are secretory granules. 600×

6. L. In addition to the short depressions or canals which lead to the apices of the parietal cells, branches occur in the depths of the gastric glands. Some argyrophil cells (ar) are located between the bases of the epithelial cells. 600×

7. L. In transverse section it becomes evident that intercellular canaliculi extend for some distance between the chief and the parietal cells. In this view the apices of the chief cells do not dominate the surface to quite the same extent as they appear to do in longitudinal sections of the gland. 600×

8. The cisternae of the granular endoplasmic reticulum in the chief cells are moderately dilated. A homogeneous content is found in these cisternae. Granules of slight, moderate, and marked density are present in the Golgi region, while those of slight to moderate density appear to reach the apex of the cell. The numerous small vesicles and some tubular profiles at the Golgi periphery have a moderately stained content. The difference in the staining reactions of the granules may indicate a difference in their content. 15,000×

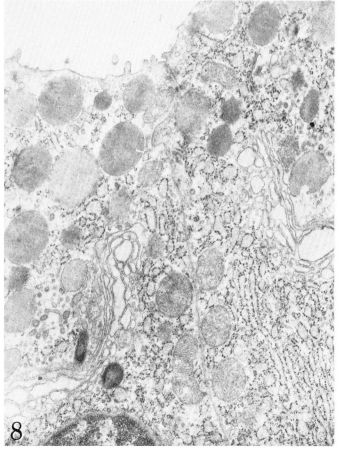

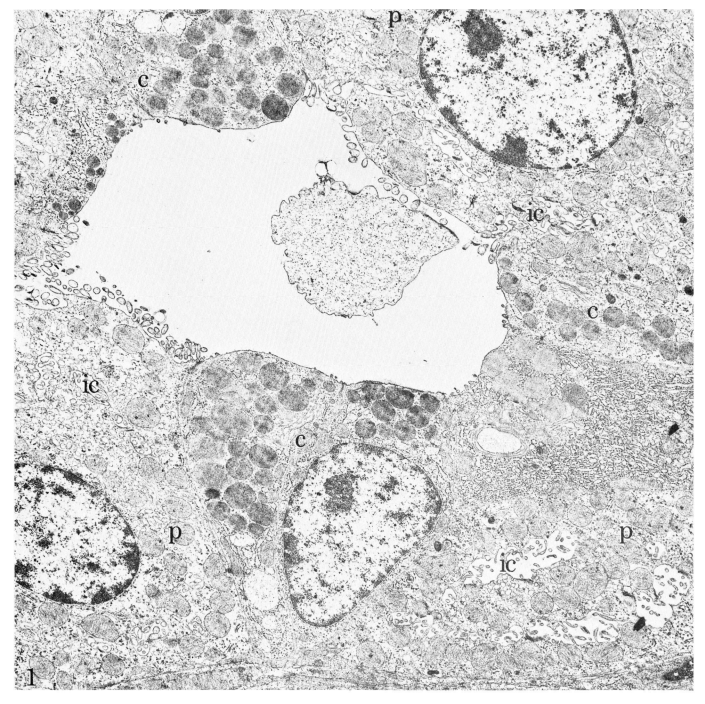

The Stomach Body

1. In transverse section of the gastric gland, the parietal cells (p) contribute to a considerable degree to the luminal surface. Microvilli are more numerous on their apices than on neighboring chief cells (c). Intracellular canaliculi (ic) open into the lumen of the gland. Numerous transverse and oblique sections through these branching intracellular canaliculi indicate the depths to which they invade the cells. Even in the depths of the cell, microvilli project into the lumen. In effect, a miniature branching tubular system similar to a compound gland exists within this cell. Small dense granules are scattered through the parietal cells. The mitochondria of the parietal cells are large, rounded, and numerous. The smooth-surfaced endoplasmic reticulum is plentiful, while the granular type is relatively scanty as compared with that of the chief cells. 7000 ×

The Duodenum

2. L. Long villi and short crypts characterize the epithelial layer of the duodenum. A layer of columnar epithelial cells in which the nuclei appear as lighter spaces constitute the surfaces of the villi. A dense, distinct content distinguishes the goblet, mucus producing cells from the absorptive cells in this epithelium. Groups of more densely stained cells, the Paneth cells, are found in the depths of the crypts. Other densely stained cells are seen in the mucous glands of Brunner, deep to the very thin muscularis mucosae. 175 ×

3. L. Duodenal villi of a young adult rat. The epithelial cells of the villi are columnar in nature with regularly placed elongated nuclei and distinct nucleoli. The cells have a striated border. This epithelium surrounds a central core in which a variety of interstitial cells, capillaries, and lymphatics (chyle channels) are

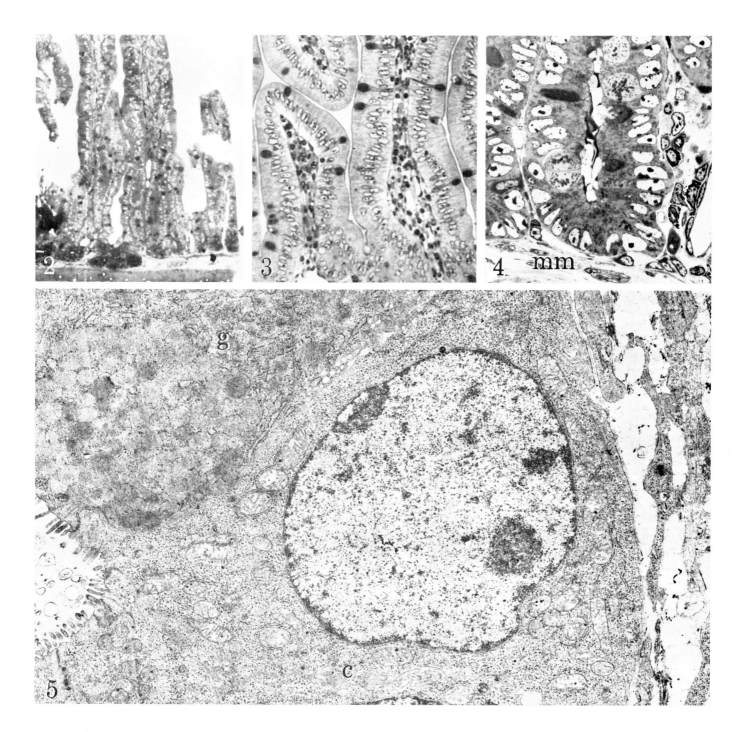

located. Goblet cells with densely stained content are interposed between the absorptive or chief cells of the epithelium. The body of the former is greatly elongated, and the cytoplasm is dense throughout. Toward the extremity of the villi the bases of the epithelial cells become quite indistinct. It is from this location that aging cells are discarded into the lumen of the intestine. 325×

4. L. A zone of very active cell division is found in the crypts of the intestine. Cells are seen in all stages of division in this area. During mitosis the bulk of the cell is drawn toward the lumen, wherein the intercellular junctional complexes appear to be retained. As mitosis is completed, the cells elongate and the daughter nuclei descend to the usual position between the other nuclei of interphase cells. It has long been known that this is the site of cell proliferation and that the aging cells are discarded

from the tips of the villi. Chyle channels, capillaries and a variety of interstitial cells in loose connective tissue, and a few smooth muscle cells, which extend from the muscularis mucosa (mm), are found between the crypts. 500×

5. The chief cell (c) of the duodenal crypt contains a slightly irregular nucleus with prominent nucleoli, a considerable number of ribosomes, some granular endoplasmic reticulum, and distinct mitochondria with pale matrices. The goblet cells (g) contain masses of what appear to be individual droplets of homogeneous material within the apices. The droplets differ in density, some being considerably darker than others. Microvilli of varying number and length are present on the luminal surface of both types of cells. The microvilli are moderately long and numerous on the cells of the upper portion of the crypt. 10,600×

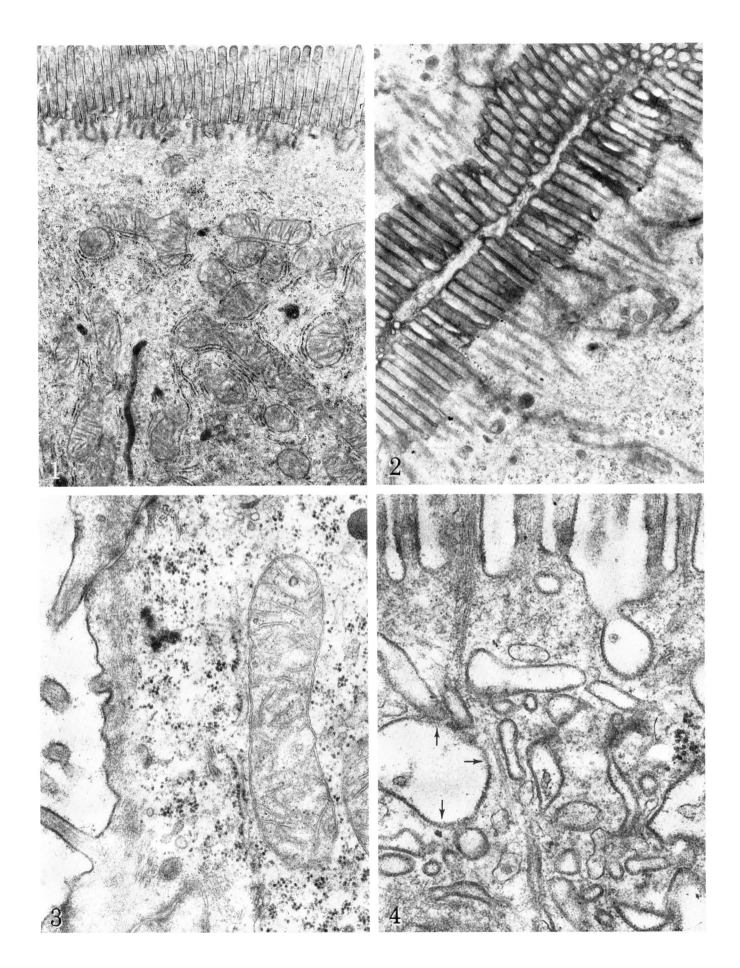

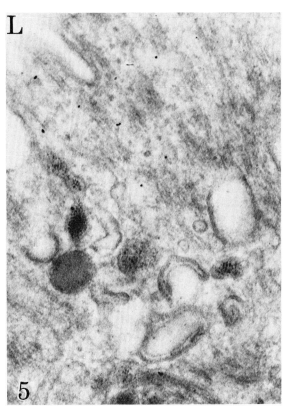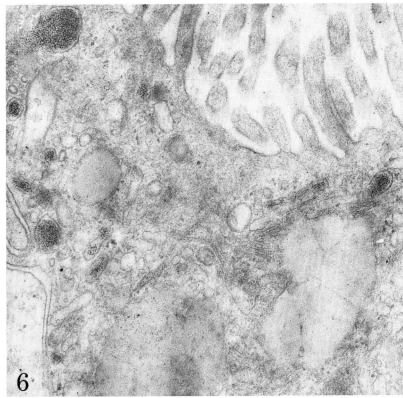

The Duodenum

1. At the base of the villi the microvilli appear tightly packed. Prominent rootlets descend some distance further into the cytoplasm from the base of these microvilli. The apex of the cell appears moderately free of organelles at this magnification. At a deeper level a number of ribosomes, granular endoplasmic reticulum, numerous mitochondria with distinct, transversely oriented crests, and some dense granules are visible. A long tube contains similar-appearing dense material. (Fixation: 6.25% glutaraldehyde with 5% DMSO in phosphate buffer followed by osmium tetroxide postfixation). 15,000×

2. At the mouth of the crypt the microvilli are long and tightly packed. Rootlets extend deep into the apex of the cell from the base of the microvilli. Some of the rootlets blend with transversely oriented filaments associated with desmosomes. Membrane-bound granules of varying density are located in the apical portion of the cell. 48,000×

3. In the depths of the crypt, microvilli on the luminal aspect of epithelial cells are shorter and widely spaced. A transverse band of filaments, the apical cell web, traverses the cell. At a slightly deeper level, mitochondria and granular endoplasmic reticulum are seen parallel to the luminal surface. A number of free ribosomes are located in this region. 55,000×

The Duodenal Epithelium—Newborn Rat

4. In this epithelium, which is actively involved in absorption, deep invaginations of the cell surface membrane are common. A network of tubes with filaments and cytoplasmic microtubules fills the apex of the cell. Where membranes have been sectioned obliquely, there is evidence of an arrangement of lines in parallel within the substructure of the membrane (arrows). 85,000×

Ileum—Newborn Rat

5. Three hours after the instillation of ferritin into the stomach of a newborn rat, the label was located in many vesicular- or tubular-appearing profiles in some epithelial cells of the small intestine. In this instance a tubule leads into the cytoplasm from the lumen (L). Dilatations in the tubule contain ferritin granules. 80,000×

6. In another instance many tubular structures and larger vesicular profiles contain ferritin. 45,000×

The intestinal epithelium of adult rats does not absorb the label. This lack of discrimination by the intestinal epithelial cells of newborn animals, which allows them to absorb colloidal gold or ferritin, lasts only a few days.

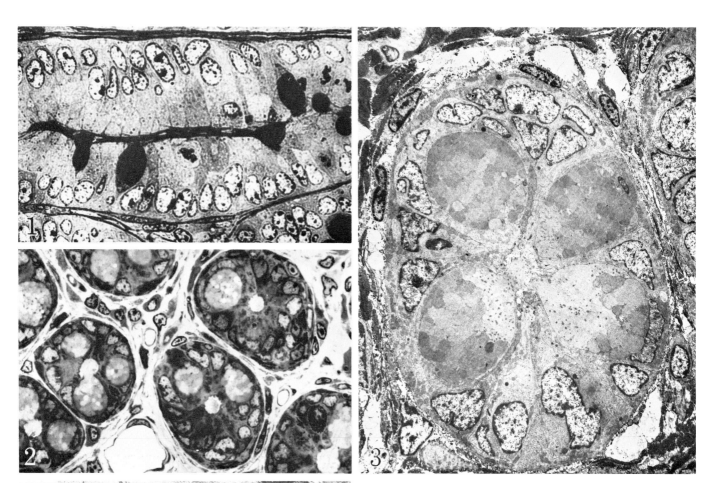

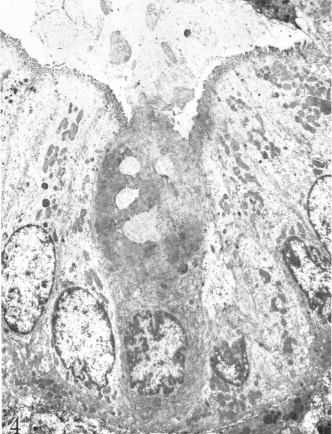

The Goblet Cells

It has been shown that protein polysaccharide complexes are assembled in the Golgi apparatus of the goblet cell.

1. L. The mucus in the goblet cells and in the lumen of the crypt of the duodenum is intensively stained by the periodic acid–silver methenamine method. In addition, terminal bars, nuclear chromatin, and nucleoli are heavily stained. The cytoplasm of the goblet cell stains more densely than that of the absorptive (chief) cell. Thus the emptied goblet cells (g) can be identified. The chromosomes of a cell in anaphase, extracellular basal laminae, and collagen stain very densely. 600×

2. L. Transversely sectioned crypts of the large intestine contain many distended goblet cells. The goblets contain more or less circular packets of mucus. Among these droplets a few dense particles can be seen. Two of the goblets are in communication with the central lumen, and it would appear that the content is in the process of being discharged into the lumen. An entero-chromaffin cell (arrow) with a few small dense cytoplasmic granules can be seen between the epithelial cells. 600×

3. As in the light micrograph, globules of varying density can be seen within the goblet in the electron micrograph. The apical content of some of the goblets has been discharged into the lumen. 2250×

4. Darker globules of mucus have been secreted from the goblet into the lumen along with a large amount of paler background material. Some pale droplets are located among the more dense

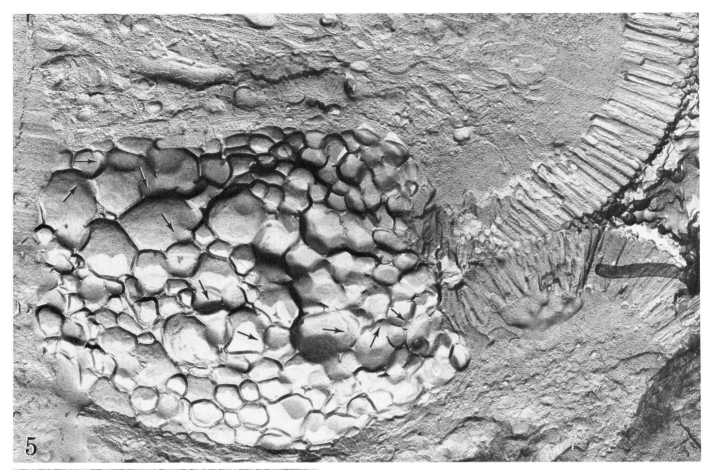

material within the goblet. A goblet cell has a large amount of granular endoplasmic reticulum surrounding its deeply indented, small, dark nucleus. The goblet cell is not as tall as, and its cytoplasm stains considerably more densely than that of the neighboring chief cells. 3500×

The Duodenum

5. In a replica of a frozen-etched specimen, the contents of a goblet cell are revealed as definite irregular globules. The surfaces of most of the globules are smooth. The goblet is compartmentalized, but most of the borders of the globules are interrupted at some point. The continuity between the compartments can be followed through the length of the goblet (arrows). Vertically oriented, elongated organelles are seen within the nearby absorptive cell. Many tubes of the endoplasmic reticulum appear to extend vertically toward the apical surface of these cells. At one point a fracture has occurred in the plane of the apical intercellular junctional complex. 14,500×

6. In an enlargement of the apical junctional complex from Fig. 5, where the fracture plane has passed between the neighboring cells, the surface of one cell has been exposed. A network of fine folds can be seen in the surface membrane within the zonula occludens, and a few circular projections extend above the surface. In the zonula adherens the surface is moderately smooth, but circular projections suggest that filaments have been fractured transversely above the membrane. In other locations the pits would suggest that filaments have been fractured at a point deep to the surface. 60,000×

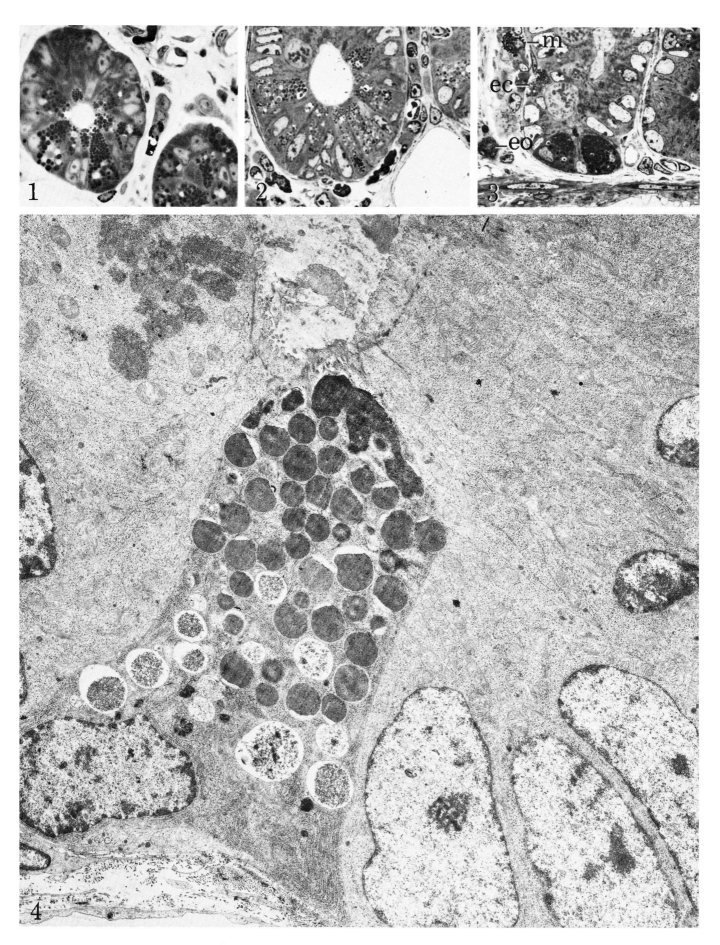

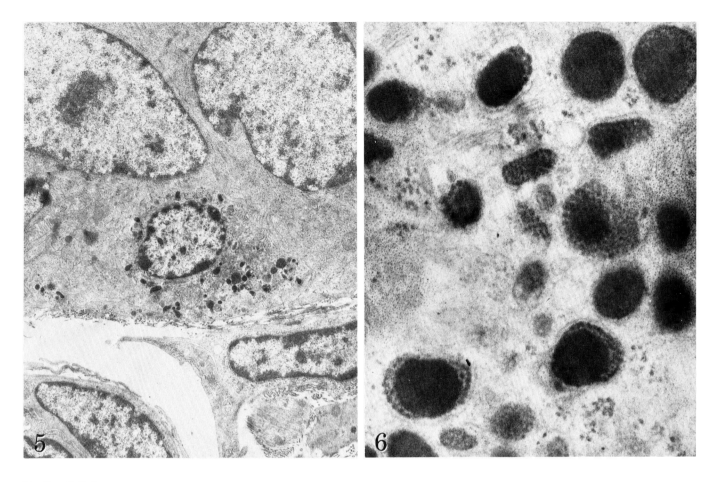

The Paneth Cells

1. L. Paneth epithelial cells found in the depths of the intestinal crypts contain numerous large dense granules. This transversely sectioned crypt is from the ileum, wherein Paneth cells are particularly numerous. The granules are seen from the base to the apex of the cells. Vacuoles are present among the granules, but this is more evident toward the base of the cells. An unusual feature of these cells is that the granules at the base are often larger and more dense than in the apex. In another crypt, the section has been obtained at a level below the lumen of the crypt resulting in transverse sections through various levels of the cells. 500×

2. L. Crypts of ileum. The base of a crypt has been sectioned slightly obliquely to the longitudinal axis of the lumen. Paneth cells occupy much of the wall of the crypt. A cell in mitosis is seen at a slightly higher level. 550×

3. L. Paneth cells are seen in the base of a longitudinally sectioned crypt, and an enterochromaffin cell (ec) is located between the epithelial cells and the basal lamina. An eosinophil (eo) and a mast cell (m) of the interstitial tissue can be distinguished by the size of their granules. The Paneth cell contains a few vacuoles as well as the dense granules. 500×

4. A Paneth cell is located among chief cells in the base of a duodenal crypt. One of the latter is in prophase. In this Paneth cell the granules increase in density toward the apex. A number of smaller, dense lysosomal granules are located among large vacuoles in a deeper level of the cytoplasm. The large vacuoles have a content of variable density. The granules closest to the apex in this cell appear to be united to form an intracellular,

irregularly shaped confluence of channels with a dense content. There is a large amount of granular endoplasmic reticulum, particularly in the base of the cell. This extends well toward the apex among the dense granules. Moderately dense material is seen in the lumen of the crypt. 7000×

Argentaffin Cell—Duodenum

It has recently been demonstrated that the argentaffin and the argyrophil cells, which were considered as representing different functional stages of the same cell, are distinct and functionally different cells.

5. The argentaffin cell is almost always seen between the bases of epithelial cells, on or near the basal lamina. This cell does not reach the intestinal lumen. It has a relatively small nucleus with fairly dense peripheral chromatin. The cytoplasm contains a number of dense granules which appear round or elongated and vary considerably in size. 6000×

6. In granules from another argentaffin cell, dense homogeneous material is superimposed upon regularly arranged circular subparticles. Some of these subparticles appear to be arranged in beadlike series. 65,000×

It has been suggested that the argentaffin cells contain 5-hydroxytryptamine (serotonin), which is a stimulant of smooth muscle.

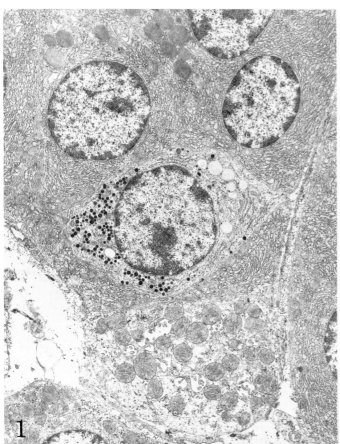

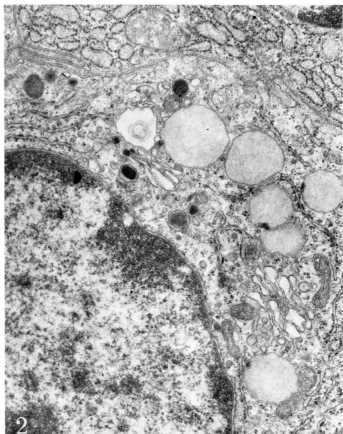

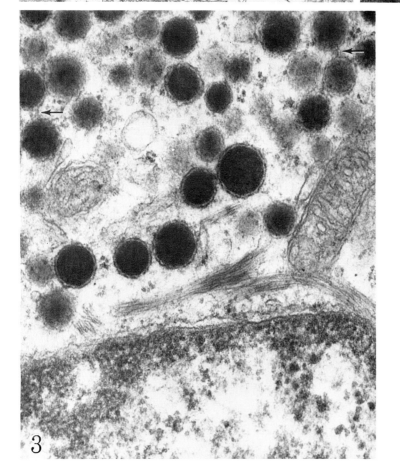

The Cholinesterase-Rich Argyrophil Cell—Stomach

These cells are most common in the pyloric region of the stomach, but are present in smaller numbers throughout the small intestine. Argyrophilia is not specific for this cell. However, it has been found to be rich in nongranular (microsomal, nonlysosomal), E 600-resistant esterase and nonspecific cholinesterase (ChE). Therefore, it has been suggested that the term "ChE-rich argyrophil" is more appropriate.

1. A ChE-rich argyrophil cell is located between the basal portion of chief cells. These cells sometimes reach the lumen of the gastrointestinal canal. The cytoplasm contains numerous round dense granules at one pole and larger vacuoles at the other. 4500 ×

2. The ChE-rich argyrophil cell contains a moderate amount of granular endoplasmic reticulum, large, smooth, membrane-bound vacuoles with very little content, and a prominent Golgi apparatus. Small dense granules are seen in the ends of some Golgi sacs. Some granules appear circular with a clear zone between the membrane and the moderate to very dense central content. 19,500 ×

3. At the other pole of the ChE-rich cell the dense granules are intermingled with mitochondria, smooth-surfaced endoplasmic reticulum, bundles of tonofilaments, and scattered ribosomes. Granular endoplasmic reticulum and microtubules are seen in the zone of cytoplasm between the nucleus and the cell border. Granules of all degrees of density in the ChE-rich cell appear to be made up of a homogeneous material. The membrane surrounding some granules has a scalloped appearance, and some granules appear to be interconnected by tubular stems (arrows). 60,000 ×

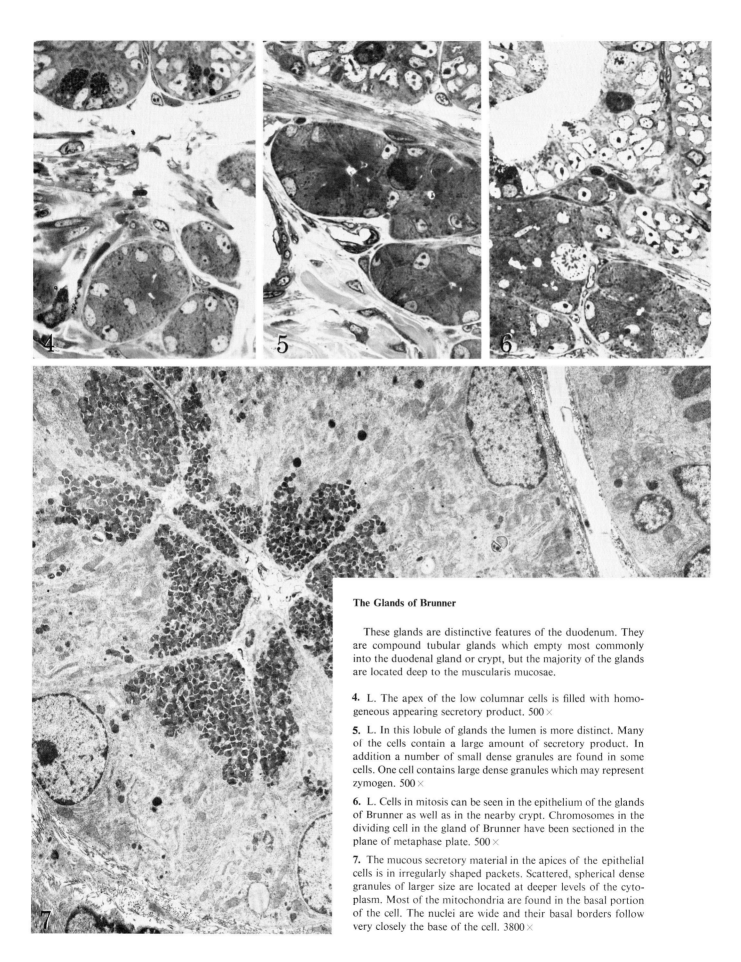

The Glands of Brunner

These glands are distinctive features of the duodenum. They are compound tubular glands which empty most commonly into the duodenal gland or crypt, but the majority of the glands are located deep to the muscularis mucosae.

4. L. The apex of the low columnar cells is filled with homogeneous appearing secretory product. 500×

5. L. In this lobule of glands the lumen is more distinct. Many of the cells contain a large amount of secretory product. In addition a number of small dense granules are found in some cells. One cell contains large dense granules which may represent zymogen. 500×

6. L. Cells in mitosis can be seen in the epithelium of the glands of Brunner as well as in the nearby crypt. Chromosomes in the dividing cell in the gland of Brunner have been sectioned in the plane of metaphase plate. 500×

7. The mucous secretory material in the apices of the epithelial cells is in irregularly shaped packets. Scattered, spherical dense granules of larger size are located at deeper levels of the cytoplasm. Most of the mitochondria are found in the basal portion of the cell. The nuclei are wide and their basal borders follow very closely the base of the cell. 3800×

REFERENCES

AMELOBLAST

Bélanger, L. F. (1955). Autoradiographic detection of radiosulfate incorporation by the growing enamel of rats and hamster. *J. Dental. Res.* **34**, 20–27.

Elwood, W. K. (1968). The ultrastructure of the enamel organ related to enamel formation. *Am. J. Anat.* **122**, 73–94.

Garant, P. R., and Nalbandian. J. (1968). Observations on the ultrastructure of ameloblasts with special reference to the Golgi complex and related components. *J. Ultrastruct. Res.* **23**, 427–443.

Kallenbach, E. (1968). Fine structure of rat incisor ameloblasts during enamel maturation. *J. Ultrastruct. Res.* **22**. 90–119.

Reith, E. J. (1968). Collagen formation in developing molar teeth of rats. *J. Ultrastruct. Res.* **21**, 383–414.

STOMACH

Cawalheira, A. F., Welsch, U., and Pearse, A. G. E. (1968). Cytochemical and ultrastructural observations on the argentaffin and argyrophil cells of the gastro-intestinal tract in mammals, and their place in the APUD series of polypeptide-secreting cells. *Histochemie* **14**, 33–46.

Helander, H. F. (1962). Ultrastructure of fundus glands of the mouse gastric mucosa. *J. Ultrastruct. Res. Suppl.* 4, 1–123.

Ito, S., and Winchester, R. J. (1963). The fine structure of the gastric mucosa in the rat. *J. Cell Biol.* **16**, 541–577.

Rubin, W., Ross, L. L., Sleisenger, M. H., and Jeffries, G. H. (1968). The normal human gastric epithelia. A fine structural study. *Lab. Invest.* **19**, No. 6, 598–626.

Stephens, R. J., and Pfeiffer, C. J. (1968). Ultrastructure of the gastric mucosa of normal laboratory ferrets. *J. Ultrastruct. Res.* **22**, 45–62.

SMALL INTESTINE

Behnke, O., and Moe, H. (1964). An electron microscope study of mature and differentiating paneth cells in the rat, especially of their endoplasmic reticulum and lysosomes. *J. Cell Biol.* **22**, 633–652.

Berlin, J. D. (1967). The localization of acid mucopolysaccharides in the Golgi complex of intestinal goblet cells. *J. Cell Biol.* **23**, 760–766.

Friend, D. (1965). The fine structure of Brunner's gland in the mouse. *J. Cell Biol.* **25**, 563–576.

Hugon, J., and Borgers, M. (1966). Submicroscopic localization of the alkaline phosphatase activity in the duodenum of the rat. *Exptl. Cell Res.* **45**, 698–702.

Hugon, J., and Borgers, M. (1967). Fine structural localization of lysosomal enzymes in the absorbing cells of the duodenal mucosa of the mouse. *J. Cell Biol.* **33**, C212–C218.

Leeson, C. R. and Leeson, T. S., (1966). The fine structure of Brunner's glands in the rat. *Anat. Record* **156**, 253–268.

Moe, H. (1968). The goblet cells, paneth cells and basal granular cells of the epithelium of the intestine. *Intern. Rev. Gen. Exptl. Zool.* **3**, 241–287.

Palay, S. L., and Karlin, L. J. (1959). An electron microscopic study of the intestinal villus. I. The fasting animal. II. The pathway of rat absorption. *J. Biophys. Biochem. Cytol.* **5**, 363–372 and 373–384.

Pearse, A. G. E., and Ricken, E. O. (1967). Histology and cytochemistry of the cells of the small intestine, in relation to absorption. *Brit. Med. Bull.* **23**, 217–222.

Rostgaard, J. and Barrnett, R. J. (1965). Fine structural observations of the absorption of lipid particles in the small intestine of the rat. *Anat. Record* **152**, 325–350.

Sjöstrand, F. S. (1967). The lipid components of the smooth-surfaced membrane-bounded vesicles of the columnar cells of the rat intestinal epithelium during fat absorption. *J. Ultrastruct. Res.* **20**, 140–160.

Staley, M. W., and Trier, J. S. (1965). Morphologic heterogeneity of mouse paneth cell granules before and after secretory stimulation. *Am. J. Anat.* **117**, 365–384.

Trier, J. S. (1963). Studies on small intestinal crypt epithelium. I. The fine structure of the crypt epithelium of the proximal small intestine of fasting humans. *J. Cell Biol.* **18**, 599–620.

Wilson, T. H. (1964). Structure and function of the intestinal absorptive cell. *In* "The Cellular Functions of Membrane Transport" (J. F. Hoffman, ed), pp. 215–229. Prentice-Hall, Englewood Cliffs, New Jersey.

Wiseman, G. (1964). "Absorption from the Intestine." Academic Press, New York.

LARGE INTESTINE

Schofield, G. C., and Silva, D. G. (1968). The fine structure of enterochromaffin cells in the mouse colon. *J. Anat. (London)* **103**, 1–13.

Silva, D. G. (1966). The fine structure of multivesicular cells with large microvilli in the epithelium of the mouse colon. *J. Ultrastruct. Res.* **16**, 693–705.

Wetzel, M. G., Wetzel, B. K., and Spicer, S. S. (1966). Ultrastructural localization of acid mucosubstances in the mouse colon with iron-containing stains. *J. Cell Biol.* **30**, 299–315.

THE MAJOR GLANDS
OF DIGESTION

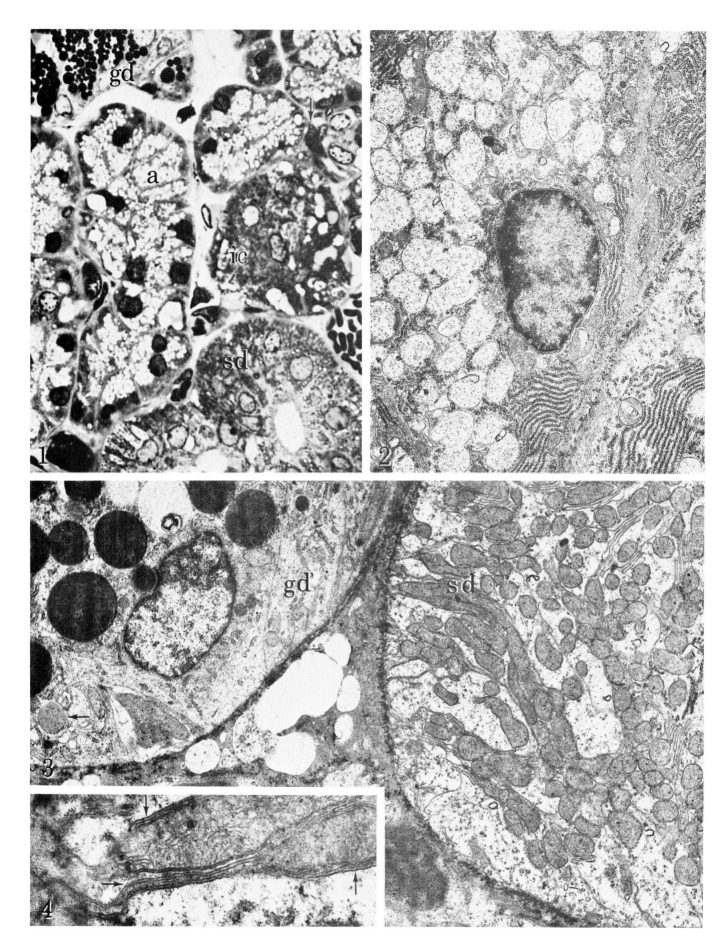

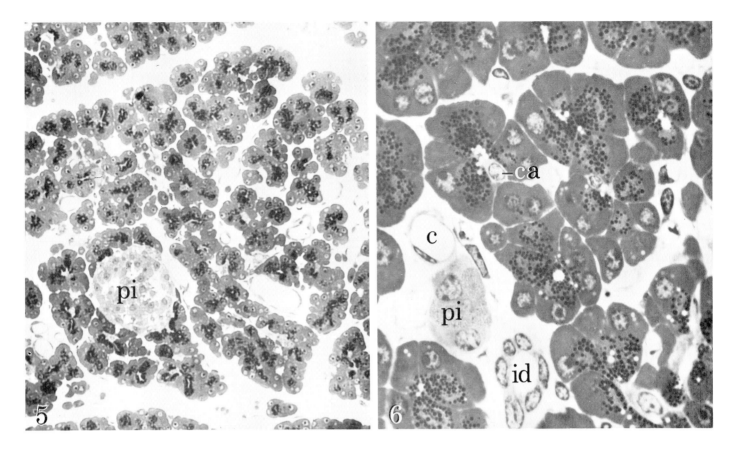

The Submandibular (Submaxillary) Gland

1. L. The submandibular gland has a variety of cells in the acini and the ducts. A narrow strip of basophilic cytoplasm is seen in the base of the cell and surrounding the small, densely stained nucleus. The acinar cell (a) is distinguished by the numerous large vacuoles which occupy most of the cytoplasm. These contain a seromucous secretion. An elongated myoepithelial cell is applied to the base of the epithelial cells in one location (arrow). The intercalated duct (ic) cells have a larger, paler nucleus and numerous mitochondria. Cells of the nonsecretory portion of the striated duct (sd) are low columnar in nature with moderately large nuclei and numerous mitochondria. The mitochondria appear to be vertically oriented, particularly in the base of the cell. In the secretory portion of the striated duct, the granulated duct (gd), the cells are filled with discrete large circular granules. The nuclei are moderately large and pale with a dense border. 700×

2. The acinar cell has a concentration of rough-surfaced endoplasmic reticulum and ribosomes in the basal cytoplasm and in a narrow band surrounding the small dark nucleus. The rest of the cell appears to be filled with very large membrane-bound vesicles, between which smaller amounts of granular reticulum can be seen. These vesicles contain dark particles superimposed upon a very light background. 6200×

3. In one type of cell from the granular duct (gd), the secretory portion of the striated duct, one finds large circular granules resembling zymogen granules. Granules near the base of the cell are smaller and, in some instances, considerably less dense (arrow). Some vesicles contain little or no dense material. The nucleus is surrounded by scattered cisternae of granular endoplasmic reticulum. In the nonsecretory portion of the striated duct (sd) the base of the cell is filled with mitochondria and a few scattered ribosomes in a pale cytoplasmic matrix. Between

the mitochondria are numerous membrane folds. These have been shown to be folds of the plasma membrane. 6900×

4. A part of Fig. 3 has been enlarged to illustrate the infoldings of the basal plasma membrane (arrows) in the nonsecreting portion of the striated duct. It has been found that digitations of neighboring cells often intrude into these folds. The mitochondria are located in the digitations. 45,000×

The Pancreas

The pancreas is a mixed gland, the major portion or exocrine pancreas being involved in the synthesis of digestive enzymes. The digestive enzymes include amylases, proteases, nucleases, and lipases. The synthesis of the enzymes takes place in acinar groups of cells, and the synthesized product is secreted into a central lumen. From the acini, the enzymes are transported to the duodenum by a system of ducts. Localized groups of endocrine cells, the pancreatic islets or islets of Langerhans, control carbohydrate metabolism. This control is affected by the secretion of insulin and glucagon into the bloodstream in response to increases or decreases in the blood sugar level.

5. L. A circular pancreatic islet (pi) made up of a collection of large pale cells is located within a lobule of acini, in which the cells appear smaller and darker. The secretion products of the exocrine pancreas appear as collections of densely stained material in the apexes of these smaller acinar cells. The cytoplasm of the acinar cells is basophilic. 160×

6. L. At higher magnification the individual zymogen granules in the acinar cells become more evident. Centroacinar (ca) cells with pale cytoplasm mark the beginning of the duct system. Two large, pale cells from the border of a pancreatic islet (pi) are located between a small intralobular duct (id) with low cuboidal epithelium and a typical capillary (c). 720×

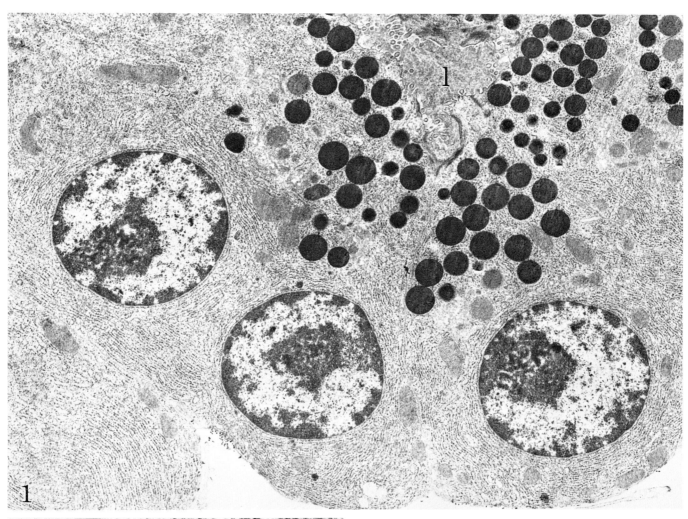

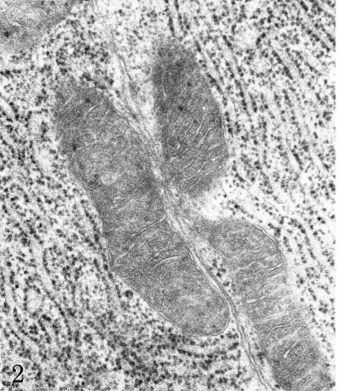

The Exocrine Pancreas

1. The nuclei of the acinar cells are small and circular with very large nucleoli. The cisterns of the granular endoplasmic reticulum occupy the cytoplasm in linear arrays from the base to the apex of the cells. A few pale vesicles are seen occasionally at the base of the cell. Granules of moderate density appear in the supranuclear region of the cell to be replaced by very dense granules toward the apex. Microvilli project into the irregularly shaped lumen (l), the content of which is of moderate density. 9200×

2. The cisternal spaces of the granular endoplasmic reticulum vary somewhat in width. Mitochondria are arranged in close proximity to the plasma membranes. Microtubules can be seen between two mitochondria. 45,000×

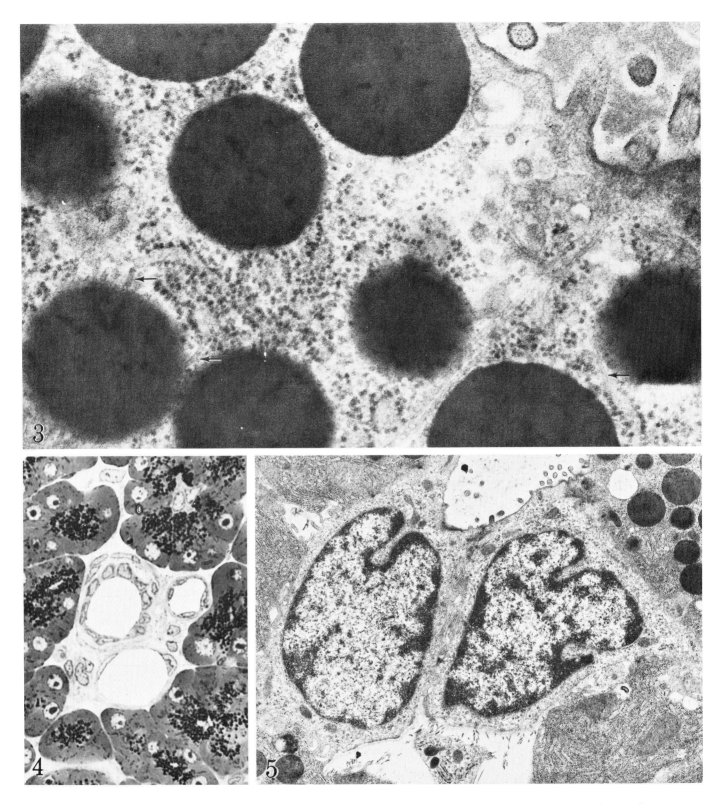

The Pancreas

3. Cytoplasmic microtubules (arrows) in the apex of a cell appear to be closely associated with zymogen granules and with smaller coated vesicles. Some of these microtubules can be followed from deep within the cell to the apical region, where they are lost in a bundle of filaments. The filaments reach the plasma membrane. In transversely sectioned microvilli the filaments in the core are seen as dense points. 70,000×

4. L. Centroacinar cells with pale cytoplasm are seen in one of the acini which surround a small duct and two accompanying blood vessels. The duct is lined by low cuboidal epithelial cells while flattened endothelial cells line the blood vessels. 600×

5. The centroacinar cells are distinguished by irregular nuclei and cytoplasm that is pale in comparison with that of the acinar cells. The cytoplasm has only a moderate supply of ribosomes. The centroacinar cells have relatively few apical microvilli. 11,250×

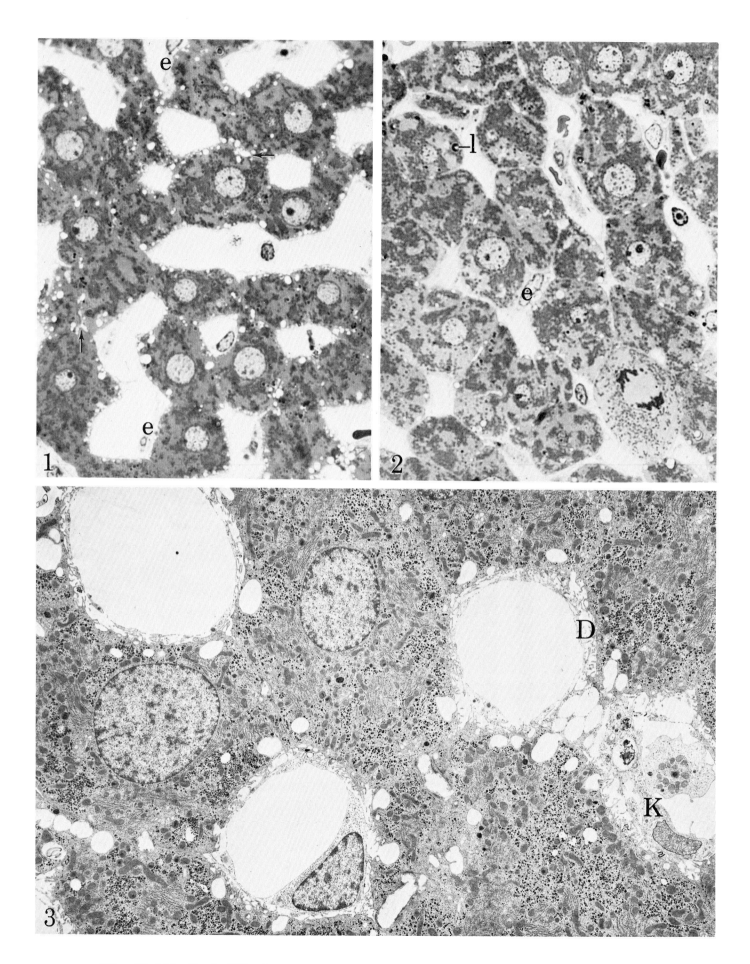

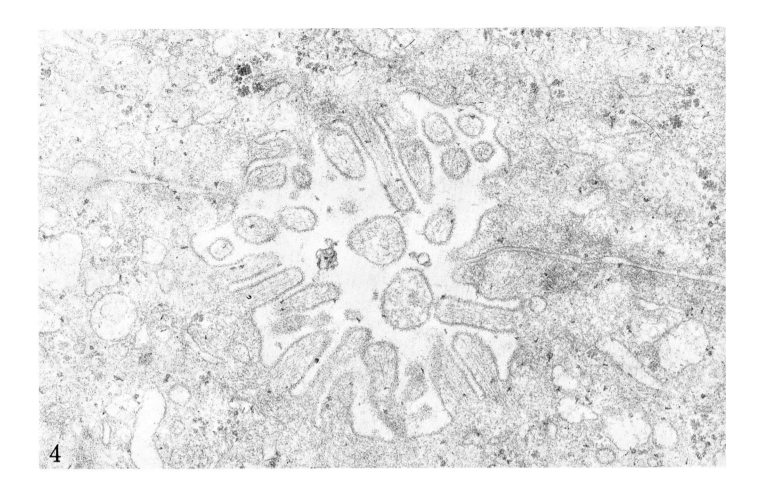

4

The Liver

In passing from the portal vein to the hepatic vein, the blood from the gastrointestinal tract and spleen flows through numerous sinusoids in the liver. The sinusoids are separated by cords of liver cells. The liver cells, under the influence of insulin from the pancreatic islets, store carbohydrates by converting glucose to glycogen. On demand, by a process of phosphorylation, it reverses this procedure, thus helping to maintain a normal blood sugar level. The liver cells perform multiple functions in the synthesis of lipids, of lipoproteins, of cholesterol, and of plasma proteins. In addition they are responsible for the metabolism and elimination of circulating hormones and drugs. A very important function of the liver is the production of bile, the salts of which are necessary for the emulsification and the absorption of fats from the intestine.

1. L. The liver parenchymal cell or hepatocyte has a large, rounded nucleus, one or more prominent nucleoli, and abundant cytoplasm. The mitochondria are darker than the background and appear as lines or masses. Very dense bodies are the lysosomes or lipofuscin granules. The sinusoidal endothelial cells (e) can be seen in widely separated locations, but the continuity of the cytoplasm necessary to complete the walls is difficult to follow with the resolution available in the light microscope. Bile canaliculi appear as irregular pores between cells in the cords (arrows). 720 ×

2. L. In other instances the mitochondria appear to occupy a greater part of the cytoplasm of the hepatocytes. Lipid droplets

(l) appear in some of the cells. Cells divide continually in the normal liver. A cell in anaphase demonstrates the location of the mitochondria in relation to the chromosomes at this stage of mitosis. Sinusoidal endothelial cells (e) in surface view appear quite large. 720 ×

3. The hepatocyte is rich in mitochondria, endoplasmic reticulum, and glycogen. Dense lysosomal granules are common in the hepatocyte. Very thin, fenestrated cytoplasmic extensions from the endothelial cells encompass the lumen of the sinusoidal blood vessel. In other areas the sinusoids are lined by a larger phagocytic cell, the stellate cell of Kupffer (K), which contains inclusions of variable density and size. Numerous cytoplasmic processes, which resemble microvilli, project from the hepatocytes into the space of Disse (D) between the liver cords and the endothelial cells. 3000 ×

4. A narrow groove of the adjacent surfaces of each of two hepatocytes contribute to an extracellular tube, the bile canaliculus. Cytoplasmic processes, the microvilli, project into the lumen. A fine fibrillar coat is visible on the external surface of the membrane of each microvillus. In addition, filamentous cores of the microvilli are seen in longitudinal and transverse section. The filaments extend into the ectoplasm of the cell as rootlets of the microvilli. These rootlets blend into a maze of filamentous or fibrillar content within the cells which encircle the canaliculus. The membranes of adjacent cells are applied more closely to one another near the lumen than at more distant locations. 60,000 ×

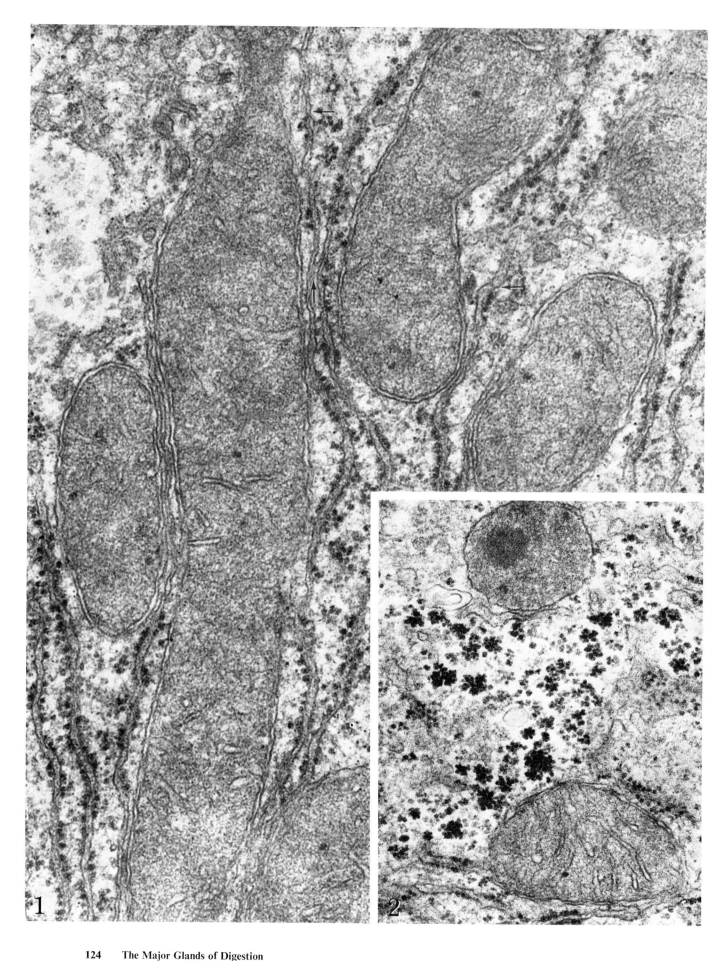

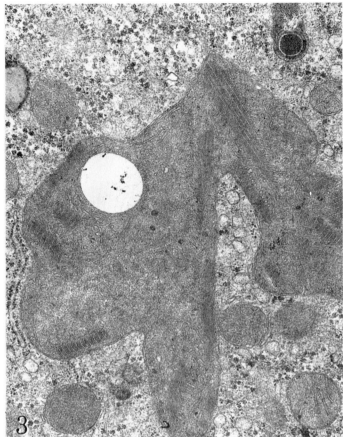

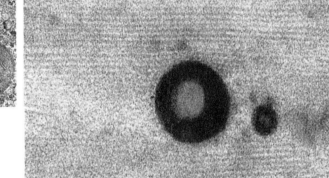

Parenchymal Cells—The Liver

1. The mitochondria of the liver have a dense matrix and transversely, or, occasionally, longitudinally oriented crests. The granular and the agranular endoplasmic reticulum are in continuity in several locations in this field (arrows). The cisternae are flattened with some content. The endoplasmic reticulum is in close association with the mitochondria. Filamentous or tubular stems (arrows) seem to interconnect some of the organelles. 90,000×

2. The microbodies generally appear somewhat smaller than the mitochondria. The microbody is surrounded by a single membrane and this opportune section has passed through a dense nucleoid. This microbody is located among a number of glycogen granules. 80,000×

The Human Liver

3. Very large, filament-bearing mitochondria are not uncommon in the normal human liver. In all of a series of eleven humans with no known liver disease, these large mitochondria were found. Their incidence varies considerably. In sections prepared for the light microscope it was found that the mitochondria which contain these paracrystalline-appearing masses of filaments were stained intensively by periodic acid–silver methanamine followed by either toluidine blue or basic fuchsin. They were found most commonly in the periportal zones. Mitochondrial crests are diminished in number, and, when present, are often arranged in series as bars or tubules along the borders of the paracrystalline masses. 27,500×

4. A transverse section of a filament-bearing mitochondrion demonstrates the arrangement of the filaments in the paracrystalline masses. A clear central zone and dark circular periphery (arrows) can be seen in filaments sectioned perfectly transversely. Large dense bodies are seen in the mitochondrion. This tissue was from a biopsy of the liver of a man acutely ill with leptospirosis icterohaemorrhagica, but the appearance of the mitochondria is probably unrelated to the pathological condition. (Fixed in 2% osmium tetroxide solution.) 95,000×

5. Large, dense bodies are seen between the filamentous masses in this mitochondrion. 105,000×

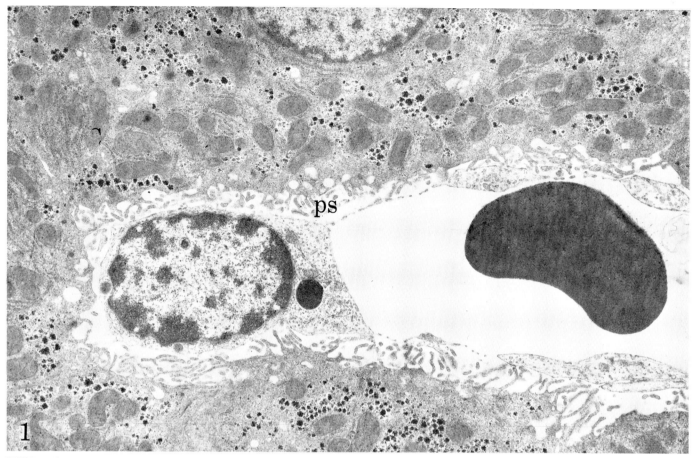

The Liver Sinusoid

1. Cytoplasmic processes from the liver parenchymal cells extend into the perisinusoidal space of Disse (ps). Thin fenestrated cytoplasm extends out from the body of the littoral cell and, alternating with thicker segments, encompasses the sinusoid. The cell contains granules of differing sizes and densities. These granular cells have been considered transitional cells between the ordinary endothelial and the Kupffer cell. A red blood cell is seen in the sinusoid. 8000×

2. In this photograph a more stellate appearance is seen in a Kupffer reticuloendothelial cell of the liver sinusoid. The body of the cell appears to occupy most of the lumen of the sinusoid. It contains a considerable number of lysosomes, a distinct Golgi complex and, in this case, a centriole surrounded by extremely densely stained satellites. A cytoplasmic microtubule and a filament extend out from one of the satellites. 17,500×

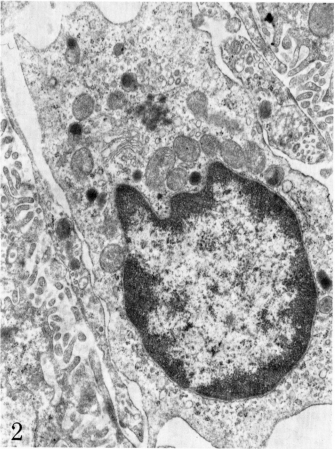

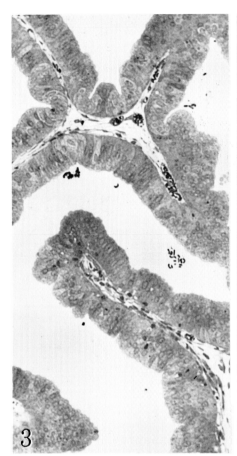

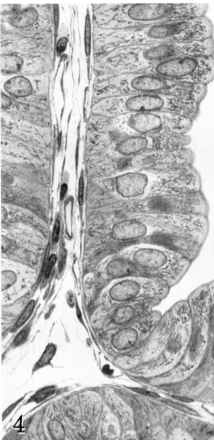

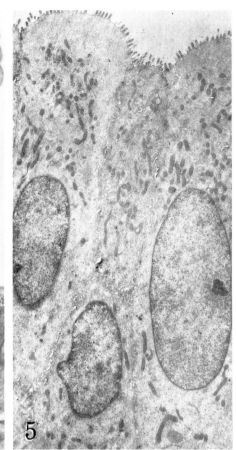

The Gallbladder—Rabbit

3. L. This section from a rabbit gallbladder demonstrates a columnar type of epithelium on an extremely folded surface. 170×

4. L. At a higher magnification the cells exhibit large ovoid nuclei, a striated border, and a considerable number of dense mitochondria against a light cytoplasmic background. Shallow pits are created in the surface of the mucosa by groups of shorter cells. In these pits the cytoplasm of the cells appears less densely stained. 680×

5. The microvilli on the apices of the epithelial cells appear short and erect. The nuclei are ovoid with some irregularity, and, in the cytoplasm, many elongated mitochondria are aligned in the vertical axis of the cell. 5200×

Bile Ducts

6. L. The common bile duct of the rat has numerous diverticulae. These probably function as reservoirs since the rat has no gallbladder. The duct is lined by columnar epithelium with a lighter cell type located between the bases of the epithelial cells. 560×

7. L. When the section does not include the neck of these divert-iculae, they present a glandular appearance. In this case two large light cells (arrow) are located adjacent to one another at the base of the epithelial cells in the diverticulum. 640×

8. The epithelial cells of the rat common bile duct have large nuclei which often have a serrated border. These cells have short microvilli on their luminal surface. In this instance the pale basal cells are embedded between the bases of the columnar cells. 9000×

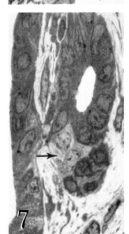

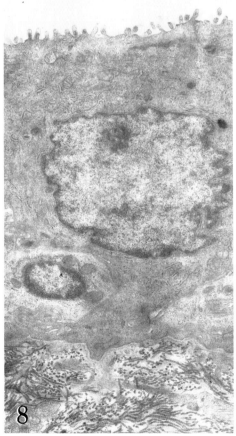

The Major Glands of Digestion   127

REFERENCES

SALIVARY GLANDS

Kurtz, S. M. (1964). The salivary glands. *In* "Electron Microscopic Anatomy" S. M. Kurtz, ed. pp. 97–122. Academic Press, New York.

Luzzatto, A. C., Procicchiani, G., and Rosati, G. (1968). Rat submaxillary gland: An electron microscope study of the secretory granules of the acinus. *J. Ultrastruct. Res.* **22**, 185–194.

Parks, H. F. (1961). On the fine structure of the parotid gland of mouse and rat. *Am. J. Anat.* **108**, 303–329.

Schneyer, L. H., and Schneyer, C. A. (1967). "Secretory Mechanisms of Salivary Glands." Academic Press, New York.

Scott, B. L., and Pease, D. C. (1959). Electron microscopy of salivary and lacrimal glands of the rat. *Am. J. Anat.* **104**, 115–161.

Sreeborg, L. M., and Meyer, J. (1964). "Salivary Glands and Their Secretions." Macmillan, New York.

PANCREAS

Caro, L. G., and Palade, G. E. (1964). Protein synthesis, storage, and discharge in the pancreatic exocrine cell. *J. Cell. Biol.* **20**, 473–495.

Ekholm, R., Zelander, T., and Edlund, Y. (1962). The ultrastructural organization of the rat exocrine pancreas. 1. Acinar cells. *J. Ultrastruct. Res.* **7**, 61–72.

Ekholm, R., Zelander, T., and Edlund, Y. (1962). The ultrastructural organization of the rat exocrine pancreas. 2. Centroacinar cells. Intercalary and intralobular ducts. *J. Ultrastruct. Res.* **7**, 73–83.

Jamieson, J. D., and Palade, G. E. (1967). Intracellular transport of secretory proteins in the pancreatic exocrine cell. II. Transport to condensing vacuoles and zymogen granules. *J. Cell Biol.* **34**, 597–615.

Palade, G. E., Siekevitz, P., and Caro, L. G. (1963). Structure, chemistry and function of the pancreatic exocrine cell. *Ciba Found. Symp., Exocrine Pancreas.* pp. 23–55.

Zelander, T., Ekholm, R., and Edlund, Y. (1962). The ultrastructural organization of the rat exocrine pancreas. 3. Intralobular vessels and nerves. *J. Ultrastruct. Res.* **7**, 84–101.

LIVER

Bruni, C., and Porter, K. R. (1965). The fine structure of the parenchymal cell of the normal rat liver. I. General observations. *Am. J. Pathol.* **46**, 691–756.

Hampton, J. C. (1964). Liver. *In* "Electron Microscopic Anatomy" (S. M. Kurtz, ed.), pp. 41–58. Academic Press, New York.

Jézéquel, A., Arakawa, K., and Steiner, J. W. (1965). The fine structure of the normal neonatal mouse liver. *Lab. Invest.* **14**, 1894–1930.

Rouiller, C., and Jezequel, A. M. (1963). Electron microscopy of the liver. *In* "The Liver," C. Rouiller (ed.), Vol. 1, p. 195. Academic Press, New York.

GALLBLADDER AND BILE DUCTS

Chapman, B. G., Chiardo, A. J., Coffey, R. J., and Weineke, K. (1966). The fine structure of the human gall bladder. *Anat. Record* **154** 579–615.

Sternlieb, I. (1965). Electron microscopic study of intrahepatic biliary ductules. *J. Microscopie* **4**, 71–80.

Yamada, E. (1955). The fine structure of the gall bladder epithelium of the mouse. *J. Biophys. Biochem. Cytol.* **1**, 445–458.

THE URINARY SYSTEM

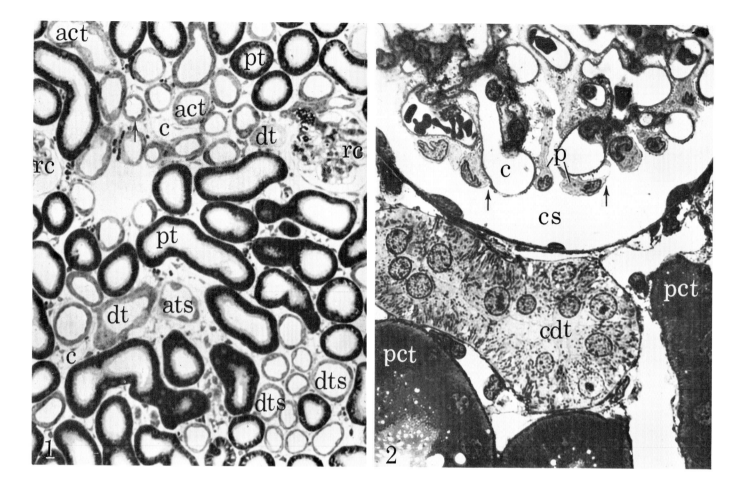

The Kidney

The kidney is made up of compound tubular glands. The portion of the tube concerned with the production of urine is called the nephron. The nephron is an unbranched tubule which empties into the branching collecting tubules.

1. L. Different segments of nephrons have been obtained in this section from the deeper part of the kidney cortex. The renal corpuscle (rc) is the dilated blind end of the nephron. In this structure there is a network of glomerular vascular channels and a labyrinthine capsular space. The vascular spaces contain a number of densely stained red blood cells. The cells of the proximal tubules (pt) have been stained intensively with toluidine blue. The tall brush borders of these cells are visible at the periphery of the lumen. Sections of the descending (dts) and thinner ascending (ats) segments of the thin-walled limb of the loop of Henle can be identified. Distal tubules (dt) and some arched collecting tubules (act) are included. An arteriole (arrow), vasa recta, and capillaries (c) are found among the tubules. 175 ×

2. L. A very thin section of the renal cortex stained with silver methenamine followed by basic fuchsin. Some of the capillaries (c) of the glomerular tuft in the renal corpuscle contain red blood cells, but most have been emptied by perfusion. The nuclei of the endothelial cells protrude into the lumina. The capsular (Bowman's) space (cs) is lined by a parietal layer of squamous epithelium and a visceral layer of much larger epithelial cells, the podocytes (p). From the podocytes, whose cell bodies are located within the capsular space, numerous cytoplasmic processes (arrows) extend toward the capillaries.

These terminate in expanded foot processes on the dense basal laminae which separate capillary and capsular epithelia. Adjacent to the corpuscle are a capillary (c) and a segment of the convoluted part of the distal tubule (cdt). At a greater distance are very densely stained segments of proximal convoluted tubules (pct). 700 ×

The Renal Corpuscle

3. The afferent arteriole breaks up into a rete mirabile, the glomerulus, in the renal corpuscles. The tortuous vessels of the rete, commonly termed glomerular capillaries, rejoin to form the efferent arteriole. The fenestrated type of endothelium forms the wall of the capillaries. In this instance some vessels contain red blood cells. The nucleus and some surrounding cytoplasm protrude into the capillary lumen. The glomerular capillaries have not been perfused in this instance, and consequently the space is greatly diminished. Most of the space is occupied by the visceral epithelial cells, the podocytes (p). Dense granules, of various sizes, can be seen in both the endothelial and in the podocyte cytoplasm. A third type of cell, which resembles the capillary pericyte, has been reported within the mesangium. These irregularly shaped mesangial cells (m) have a more densely stained cytoplasm than that of the capillary endothelium (ce) and appear to contain a higher number of dense granules. The mesangial cells are surrounded by a wide zone of dense intercellular matrix which may be related to the basal lamina. It is not known whether these are modified endothelial or smooth muscle cells. The mesangial cells may support the capillary loops and, in addition, are phagocytic. 3400 ×

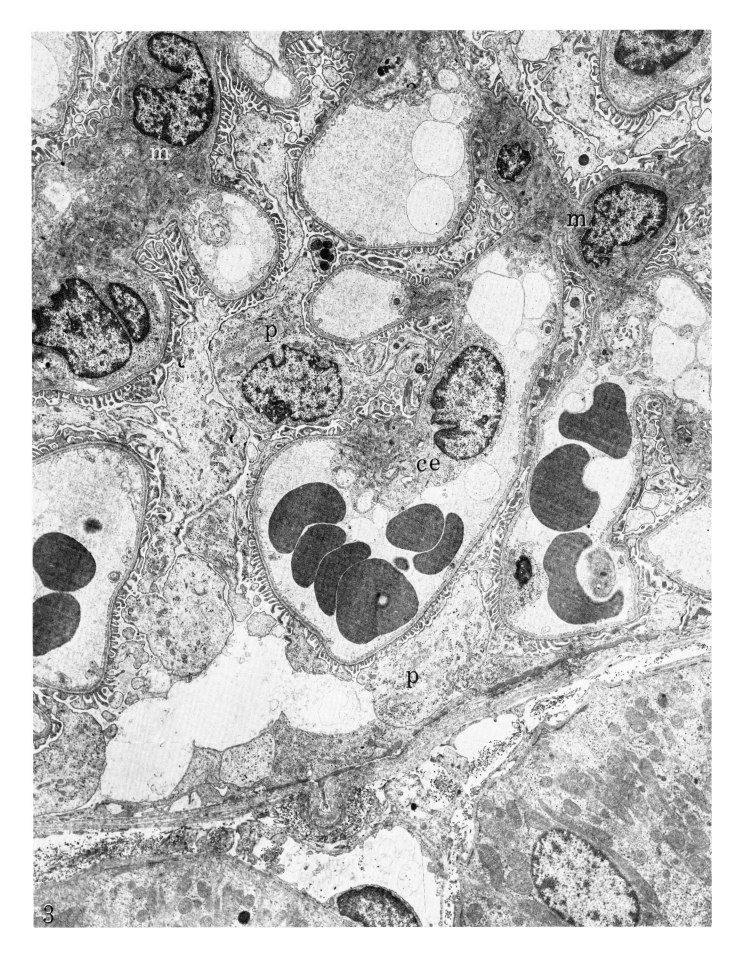

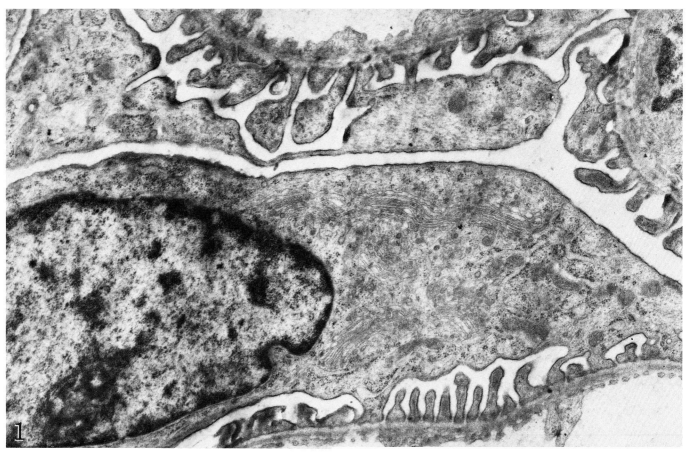

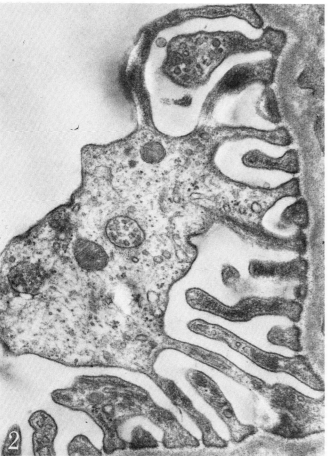

The Renal Corpuscle

1. The body of the podocyte contains a very large Golgi apparatus, a number of ribosomes, and some granular endoplasmic reticulum. Cytoplasmic processes, only a small percentage of which can be obtained in continuity in any one section, extend out to the terminal expansions or pedicels. The processes vary in complexity, in length, and in thickness. The expansions rest on a well-developed basal lamina, which is shared by the endothelial cells. This basal lamina has been considered as the main barrier to the passage of larger molecules in the process of filtration which takes place from the capillaries into the capsular space. However, recent work suggests that the main barrier may be the membrane-like sheet which spans the slit between the foot plates. The capillary endothelium is fenestrated. 27,500 ×

2. In this podocyte an unusually high number of foot processes have been obtained in continuity. The cytoplasm contains a multivesicular body, agranular endoplasmic reticulum, mitochondria, and microtubules which extend out into the foot processes. A diaphragm bridges the slit between the pedicels. Since this is a continuous diaphragm between the pedicels, the visceral layer of Bowman's capsule is completed by it. 40,000 ×

The Juxtaglomerular Apparatus

Granulated cells in the muscular layer of the afferent arteriole are the distinguishing features of the juxtaglomerular apparatus. This arrangement is located just where the arteriole enters the renal corpuscle. Recent evidence indicates that these granules contain renin, which is a vasopressor.

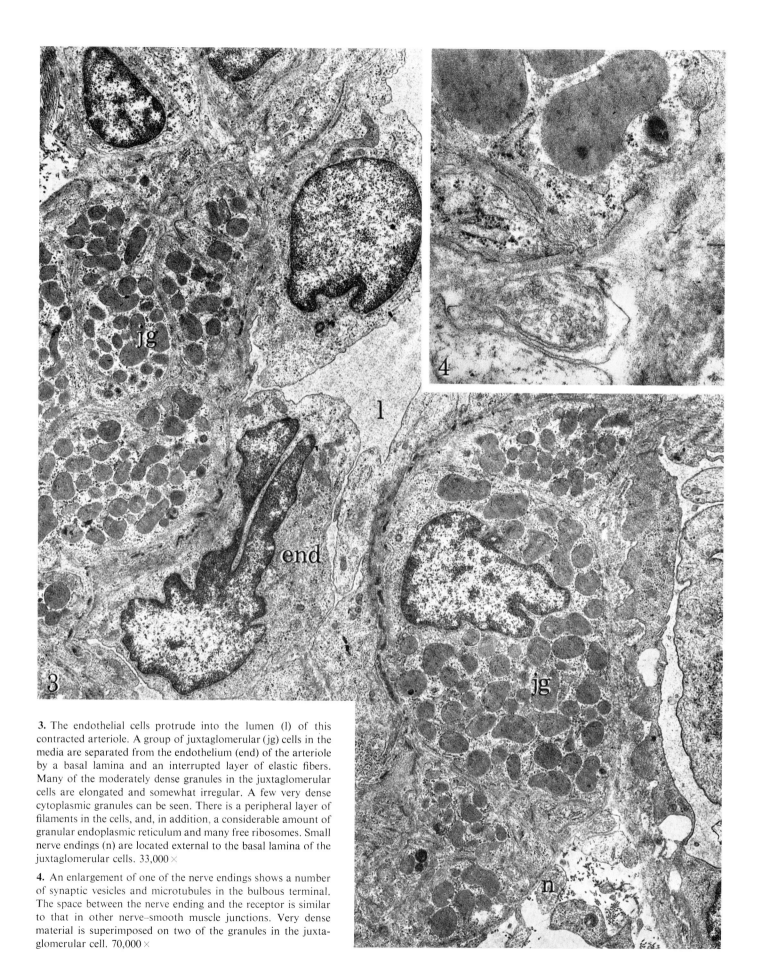

3. The endothelial cells protrude into the lumen (l) of this contracted arteriole. A group of juxtaglomerular (jg) cells in the media are separated from the endothelium (end) of the arteriole by a basal lamina and an interrupted layer of elastic fibers. Many of the moderately dense granules in the juxtaglomerular cells are elongated and somewhat irregular. A few very dense cytoplasmic granules can be seen. There is a peripheral layer of filaments in the cells, and, in addition, a considerable amount of granular endoplasmic reticulum and many free ribosomes. Small nerve endings (n) are located external to the basal lamina of the juxtaglomerular cells. 33,000 ×

4. An enlargement of one of the nerve endings shows a number of synaptic vesicles and microtubules in the bulbous terminal. The space between the nerve ending and the receptor is similar to that in other nerve–smooth muscle junctions. Very dense material is superimposed on two of the granules in the juxtaglomerular cell. 70,000 ×

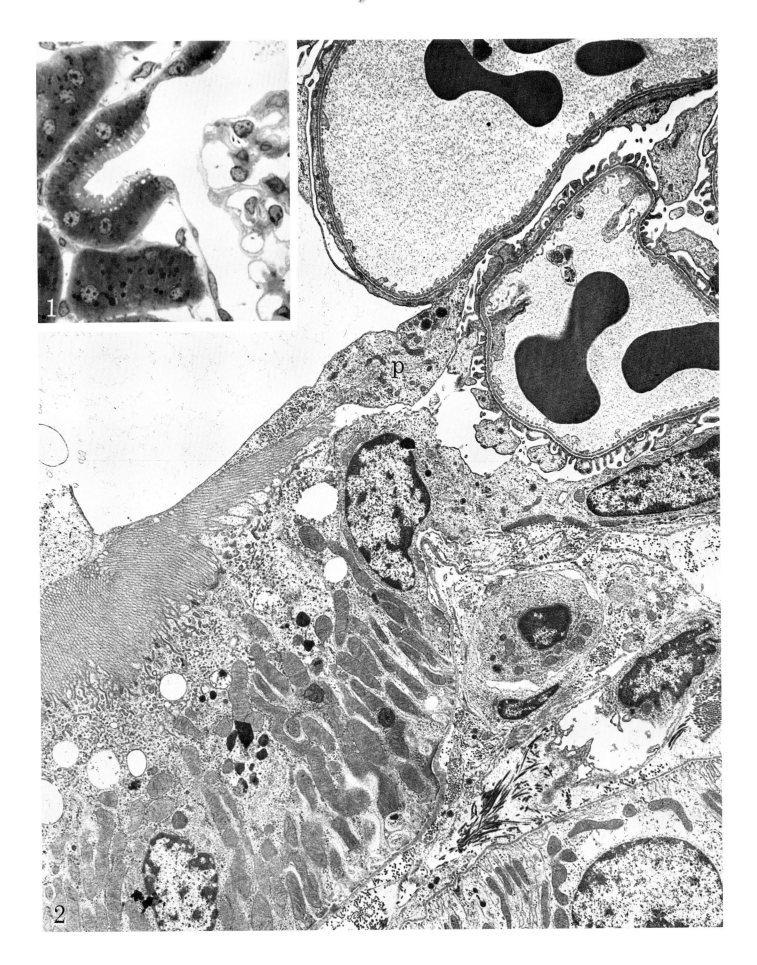

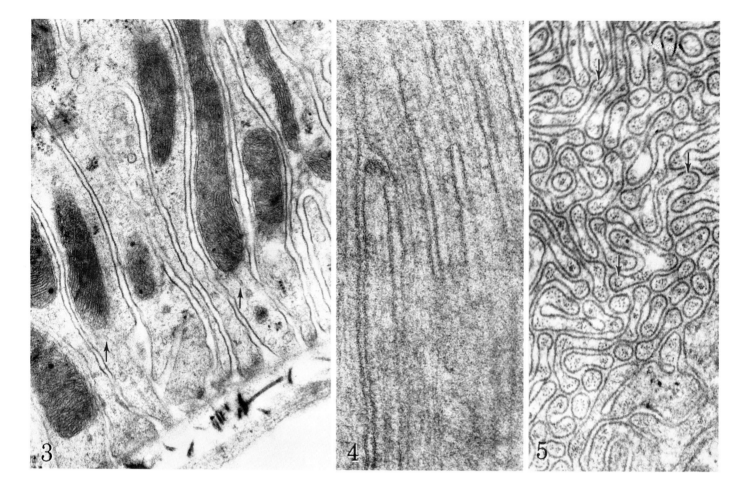

The Urinary System

After leaving Bowman's capsule the filtrate is passed through the tubules of the proximal and the distal nephron, during which time the content is constantly altered before it reaches the excretory passages of the kidney.

1. L. The parietal layer of Bowman's capsule of the renal corpuscle is continuous with the proximal tubule of the nephron. There is an abrupt change from squamous epithelium to cuboidal. The epithelial cell of the proximal convoluted tubule has a rounded nucleus, dense cytoplasm, and a tall brush border. These cells contain a number of dense cytoplasmic granules. Large, dense granules are seen in the cytoplasm of other epithelial cells of the same nature which have been sectioned obliquely and tantentially. 500×

2. At the zone of transition from the renal corpuscle to the proximal convoluted tubule the last epithelial cell of the parietal layer of Bowman's capsule has become considerably thickened and a number of dense granules have appeared. Long, thin microvilli make their appearance on the apices of cuboidal epithelial cells of the proximal convoluted tubule. A foot process of a podocyte (p) within the capsular space is in contact with the tips of the long apical microvilli of the first proximal tubule cell. Mitochondria are oriented vertically in the epithelial cell of the proximal convoluted tubule. A number of lysosomal dense granules are seen throughout the cytoplasm, and numerous, densely stained tubules occupy the cell apices. Varying amount

of dense content are superimposed on a clear background in some of the large circular vesicles in the apex of the cell. 7000×

The Proximal Tubule

3. The layers of plasma membrane which extend from the base of the cell up between the mitochondria has been interpreted as presenting either infolding of this basal surface of a single cell or zones of contact between interdigitations of folded borders of adjacent cells. Interdigitations occur frequently. At the base of the cell the mitochondria in the cytoplasmic compartments have transversely or longitudinally oriented crests. Cytoplasmic microtubules share the compartments with the mitochondria and often extend toward the base of the cell (arrows). 50,000×

4. A small number of centrally located filaments are found in each microvillus on the apex of a proximal tubule epithelial cell. A trilaminar plasma membrane encompasses each microvillus, and a very fine external coat of fibrillar material can be seen. 120,000×

5. A section has passed transversely through the apex of a cell and through microvilli which project into the lumen of a proximal tubule. Numerous cytoplasmic projections at the base of the microvilli have been sectioned transversely, thus obtaining filamentous rootlets of several microvilli within the same fold. A common number of filaments in the microvillus or its rootlet is nine. In several instances the filaments are arranged in circles (arrows). 70,000×

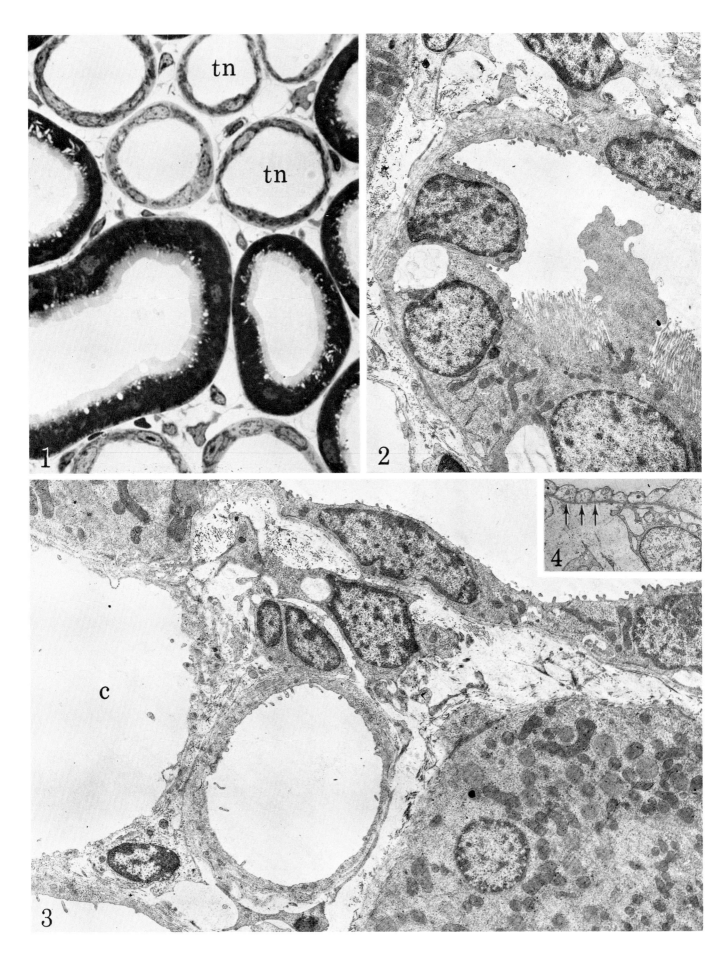

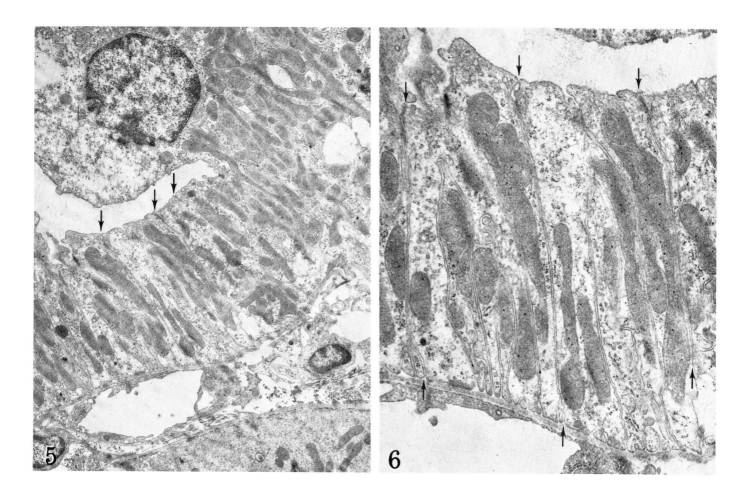

The Proximal Nephron

1. L. The wall of the straight portion of the proximal tubule with its tall brush border diminishes somewhat in thickness before giving way abruptly to the thin-walled loop of Henle. The descending, thin-walled section (tn) of the loop of Henle is made up of squamous epithelium. The mitochondria and the background cytoplasm of the cells of this part of the loop retain a moderate density. 700 ×

2. At the point of transition from the straight portion of the proximal tubule to the thin-walled loop of Henle the shape of the cells changes abruptly from cuboidal to squamous, and the lysosomes diminish in number. During the transition there is a sudden change from the numerous long, thin apical microvilli to widely separated short, thick microvillous projections. There is a rupture in the apex of the cell which is probably an artifact. 4500 ×

3. Kidney medulla. The short apical microvilli on the squamous epithelium persist through the descending part of the thin-walled loop. The thin-walled tube is closely associated with fenestrated capillaries (c). A number of interstitial cells are located between the tubules. These cells have long cytoplasmic processes which include some dense granules and resemble macrophages. Water and sodium are said to diffuse readily across the epithelium of the descending thin-walled loop. 5250 ×

4. Papilla. In the short ascending limb of the thin-walled loop of Henle, cytoplasmic processes of the squamous epithelium of neighboring cells interdigitate with one another. In transverse section, these interdigitations appear as a series of isolated cytoplasmic units (arrows). Interstitial cells with long cytoplasmic processes are located between the tubules. Active pumping of sodium out of the nephron takes place in the ascending thin limb of the loop of Henle. 3500 ×

The Medulla of the Kidney

5. In the zone of transition from the ascending thin limb to the thick limb of the loop of Henle, there is an abrupt increase in the height of the epithelial cells and the nucleus is seen in the apical region. At this point the interdigitating cytoplasmic processes become taller and appear quadrangular in transverse section. The vertically oriented mitochondria stretch from the base to the apex of what appear as separate cellular compartments. The double layer of plasma membranes extends from the apical terminal bars (arrows) to the base of the cells. A small number of dense granules are seen in the cytoplasm. 4400 ×

6. An enlargement of a part of Fig. 1 illustrates the interdigitations which are outlined by plasma membrane partitions from base to apex of the cells (arrows). Mitochondria almost reach the two poles of some of these compartments. The cells rest on a basal lamina which is shared in part by the fenestrated endothelium of a capillary. 11,000 ×

The Urinary System 137

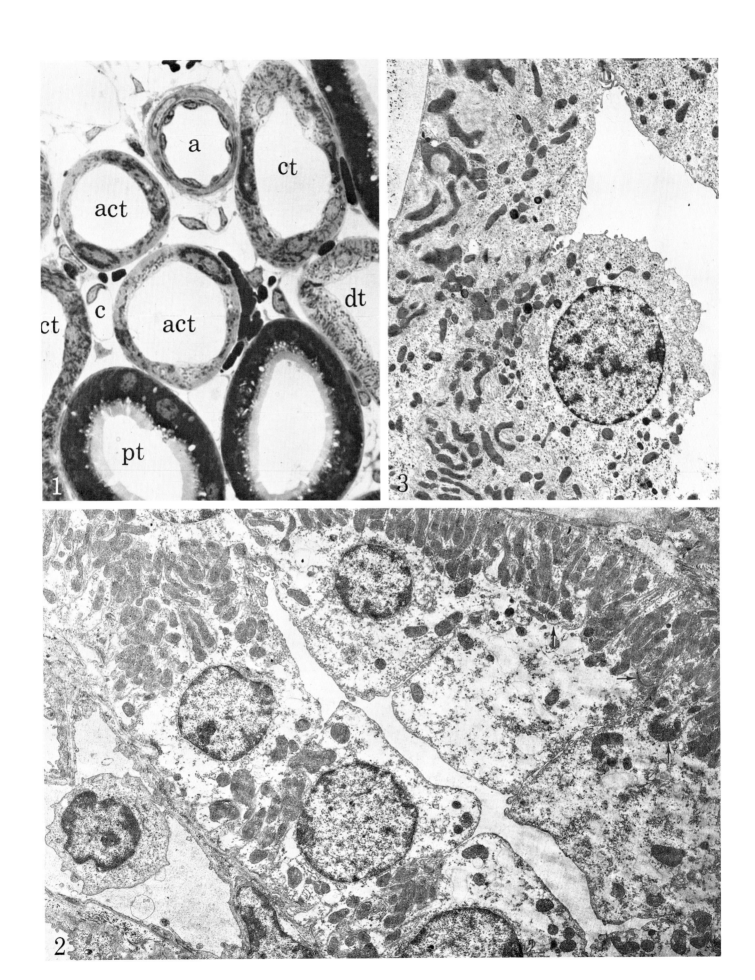

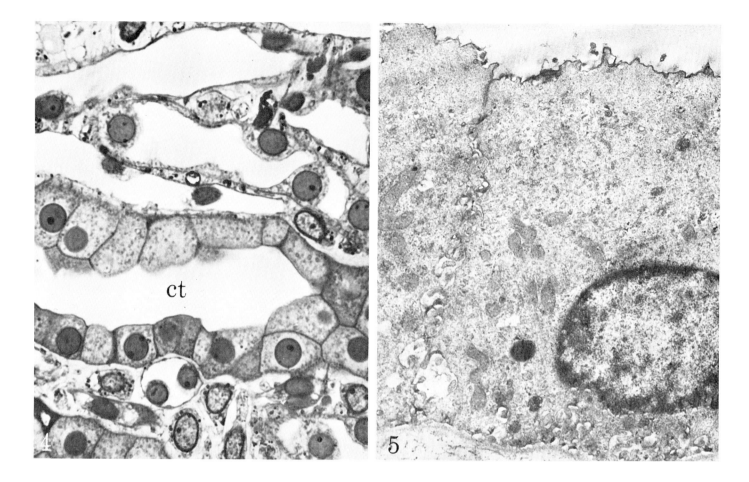

The Kidney

1. L. Proximal tubules (pt), a distal tubule (dt), connecting tubules (ct) between the distal nephron and the arched collecting tubules (act), an arteriole (a), and capillaries (c) have been obtained in one field. In the transitional zone between the distal nephron and the collecting tubules, a marked reduction in the height of the cells occurs. In the connecting portion (ct) the nuclei are still mainly in the apices of the cell, but the mitochondria become shorter, lose their basal arrangement, and are distributed throughout the cytoplasm. By utilizing a cobalt–silver impregnation, dark and light cells have been reported in the walls of the arched collecting tubules. These are quite evident when stained with toluidine blue and differentiated in the method employed for this work. The stain of the dark cells is due partly to "granulated mitochondria." 700 ×

2. In a collapsed section of the distal convoluted tubule the apical protrusions containing the nucleus appear rectangular. Consequently the apical terminal bar appears to be some distance below the luminal surface. Most of the short microvilli are in the clefts between the apical protrusions. Folds of plasma membrane or interdigitations extend between the mitochondria approximately one-half the distance toward the apparent apex of the cell, but reach the level of the apical intercellular junctional complexes in other instances. Folds of the plasma membrane loop over the mitochondria and a terminal bar can be seen at the apex of one of these loops (arrows). This is

evidence that these are interdigitations of neighboring cells. A capillary adjacent to the base of the epithelial cells contains a medium-sized lymphocyte. 5500 ×

3. The cells decrease in height in the connecting portion between the distal nephron and the small, arched collecting tubules. The mitochondria are not as regularly arranged in the base of the cell, and a larger number appear in the apical portion. Microvilli become slightly longer and more numerous. Dense granules are found in increasing numbers in these cells. 5500 ×

The Renal Papilla

4. L. The cells of the larger collecting tubule (ct) are cuboidal and have small mitochondria scattered throughout their cytoplasm. A considerable number of irregularly shaped interstitial cells with many cytoplasmic dense granules are found between the collecting tubules, the ascending part of the thin-walled loops of Henle, and the capillaries. 700 ×

5. The epithelial cells of the larger collecting tubules are cuboidal in shape with an irregular distribution of short, thick apical microvilli. The nuclei are moderately rounded with fairly dense chromatin. A number of granules of varying density but a low content of other organelles are found in the cytoplasm. The lateral borders of the cells are quite folded. 9000 ×

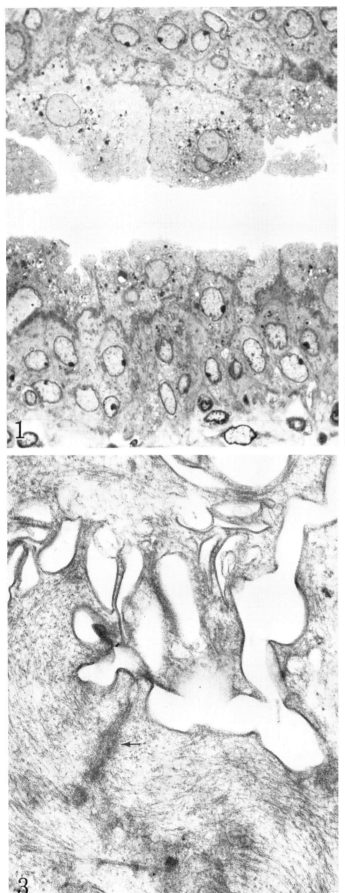

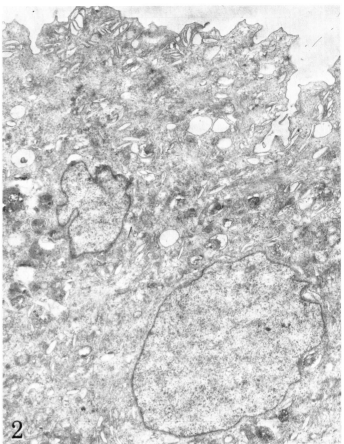

The Bladder

Beyond the papilla the epithelium of the larger conducting passages of the urinary system is stratified and is termed transitional epithelium.

1. L. There is a marked variation in the size of the cells in the transitional epithelium of the urinary bladder. The most superficial layer is made up of cells with a great amount of vacuolated cytoplasm. Dense granules increase in number and size from the middle layers of cells to those toward the luminal surface. The large vacuolated cells have deep indentations in their surfaces and the cells themselves protrude into the lumen. 720 ×

2. The vacuolation of the free surfaces of the superficial cells can be seen to be produced by very deep, branching invaginations of the surface membrane. The nuclei of the cells are large and somewhat irregular, but they have a homogeneous content. 6000 ×

3. The apices of the most superficial cells of the bladder epithelium bulge into the bladder lumen. The apical junctional complexes (arrow) are found in the depths of clefts between the cells. The junctional complex is associated with a large number of filaments within the cytoplasm. 45,000 ×

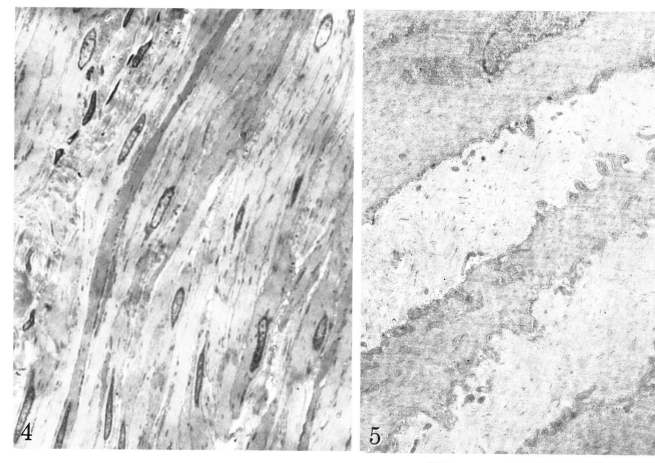

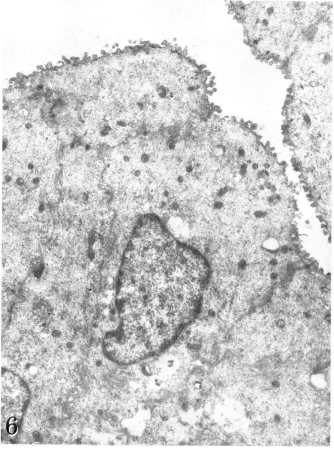

4. L. A distinct variation in cytoplasmic density is seen in the smooth muscle cells of the urinary bladder. We have found this feature in the smooth muscle of other organs, but it is most striking in the urinary bladder. The length of a stretched smooth muscle fiber is indicated by segments of fibers included in this photograph. 800×

5. In electron microscopy the difference in density between the light and the dark types of smooth muscle fiber is as evident as in light microscopy. The mitochondria appear larger and more dense in the lighter type of fiber. 3600×

The Penile Uretha

In the corpus spongiosum the penile uretha is seen as a collapsed tube surrounded by vessels and large sinusoids in a very loose connective tissue. The mucosa of the uretha is made up of the transitional type of epithelium through most of its length.

6. The superficial layers of epithelial cells lining the uretha have irregularly shaped nuclei and a few scattered mitochondria, granules, and vesicles in the cytoplasm. Projecting into the lumen is a regular array of short stout microvilli. 7000×

REFERENCES

URINARY BLADDER

Bartoszewicz, W., and Barrnett, R. J. (1964). Fine structural localization of nucleoside phosphatase activity in the urinary bladder of the toad. *J. Ultrastruct. Res.* **10**, 599–609.

El-Badawi, A., and Schenk, A. (1966). Dual innervation of the mammalian urinary bladder. A histochemical study of the distribution of cholinergic and adrenergic nerves. *Am. J. Anat.* **119**, 405–428.

KIDNEY

Barajas, L. (1964). The innervation of the juxtaglomerular apparatus. An electron microscopic study of the innervation of the glomerular arterioles. *Lab. Invest.* **13**, 916–929.

Barajas, L. (1966). The development and ultrastructure of the juxtaglomerular cell granule. *J. Ultrastruct. Res.* **15**, 400–413.

Barajas, L., and Latta, H. (1963). A three-dimensional study of the juxtaglomerular apparatus in the rat. *Lab. Invest.* **12**, 257–269.

Bulgar, R. E., Tisher, C. C., Myers, C. H., and Trump, B. F. (1967). Human renal ultrastructure. II. The thin limb of Henle's loop and the interstitum in healthy individuals. *Lab. Invest.* **16**, 124–141.

Edelman, R., and Hartroft, P. M. (1961). Localization of renin in the juxtaglomerular cells of the rabbit and dog through the use of the fluorescent antibody technique. *Circulation Res.* **9**, 1069–1077.

Graham, R. C., Jr., and Karnovsky, M. J. (1966). The early stages of absorption of injected horseradish peroxidase in the proximal tubules of mouse kidney: Ultrastructural cytochemistry by a new technique. *J. Histochem. Cytochem.* **14**, 153–164.

Hatt, P. Y. (1967). The juxtaglomerular apparatus. *In* "Ultrastructure of the Kidney" (A. J. Dalton, and F. Haguenau, eds.), pp. 101–141. Academic Press, New York.

Hicks, R. M. (1965). The fine structure of the transitional epithelium of rat ureter. *J. Cell Biol.* **26**, 25—48.

Johnson, F. R., and Darnton, S. J. (1967). Ultrastructural observations on the renal papilla of the rabbit. *Z. Zellforsch. Mikroskop. Anat.* **81**, 390–406.

Kurtz, S. M. (1964). The kidney. *In* "Electron Microscopic Anatomy" (S. M. Kurtz, ed.), pp. 239–265. Academic Press, New York.

Latta, H. (1961). Cilia in different segments of the rat nephron. *J. Biophys. Biochem. Cytol.* **11**, 248–252.

Latta, H., Maunsbach, A. B., and Osvaldo, L. (1967). The fine structure of renal tubules in cortex and medulla. *In* "Ultra-structure of the Kidney". (A. J. Dalton, and F. Haguenau, eds.), pp. 1–56. Academic Press, New York.

Latta, H., Stone, R. S., Bencosme, S. A., and Madden, S. C. (1961). Mechanisms for movement of fluid in renal tubule cells. *In* "Electron Microscopy in Anatomy" (Anat. Soc. Gt. Brit. and Ireland, ed.), pp. 235–270. Williams & Wilkins, Baltimore, Maryland.

Maunsbach, A. B. (1966). Absorption of I^{125}-labeled homologous albumin by rat kidney proximal tubule cells. A study of microperfused single proximal tubules by electron microscopic autoradiography and histochemistry. *J. Ultrastruct. Res.* **15**, 197–241.

Menefee, M. G., and Mueller, C. B., (1967). Some morphological considerations of transport in the glomerulus. *In* "Ultrastructure of the Kidney" (A. J. Dalton, and F. Haguenau, eds.), pp. 73–100. Academic Press, New York.

Michielsen, P., and Creemers J. (1967). The structure and function of the glomerular mesangium. *In* "Ultrastructure of the Kidney" (A. J. Dalton, and F. Haguenau, eds.), pp. 57–72. Academic Press, New York.

Miller, F., and Palade, G. (1964). Lytic activities in renal protein absorption droplets. An electron microscopical cytochemical study. *J. Cell Biol.* **23**, 519–552.

Neustein, H. B. (1967). Hemoglobin absorption in the proximal tubules of the kidney in the rabbit. *J. Ultrastruct. Res.* **17**, 565–587.

Neustein, H. B., and Maunsbach, A. B. (1966). Hemoglobin absorption by proximal tubule cells of the rabbit kidney. A study of electron microscopic autoradiography. *J. Ultrastruct. Res.* **16**, 141—157.

Osvaldo, L., and Latta, H. (1966). The thin limb of the loop of Henle. *J. Ultrastruct. Res.* **15**, 144–168.

Osvaldo, L., and Latta, H. (1966). Interstitial cells of the renal medulla. *J. Ultrastruct. Res.* **15**, 589–613.

Patrizi, G., and Middelkamp, J. N. (1967). The distribution and pattern of the agranular reticulum in rat kidney tubules. *J. Microscopie* **6**, 91–94.

Pease, D. C. (1955). Electron microscopy of the tubule cells of the kidney cortex. *Anat. Record* **121**, 723–743.

Pease, D. C. (1955). Fine structures of the kidney seen by electron microscopy. *J. Histochem. Cytochem.* **3**, 295–308.

Rhodin, J. (1958). Anatomy of kidney tubules. *Intern. Rev. Cytol.* **4**, 485–534.

Tischer, C. C., Bulger, R. E., and Trump, B. F. (1966). Human renal ultrastructure. I. Proximal tubule of healthy individuals. *Lab. Invest.* **15**, 1357–1394.

Yamada, E. (1955). The fine structure of the renal glomerulus of the mouse. *J. Biophys. Biochem. Cytol.* **1**, 551–566.

THE MALE
REPRODUCTIVE
SYSTEM

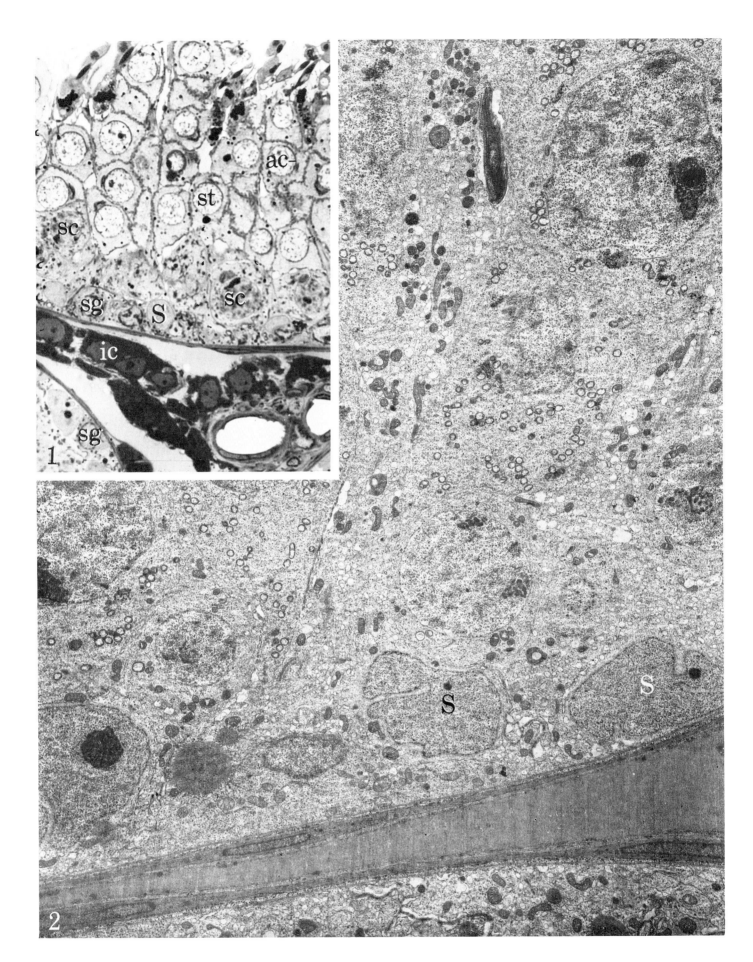

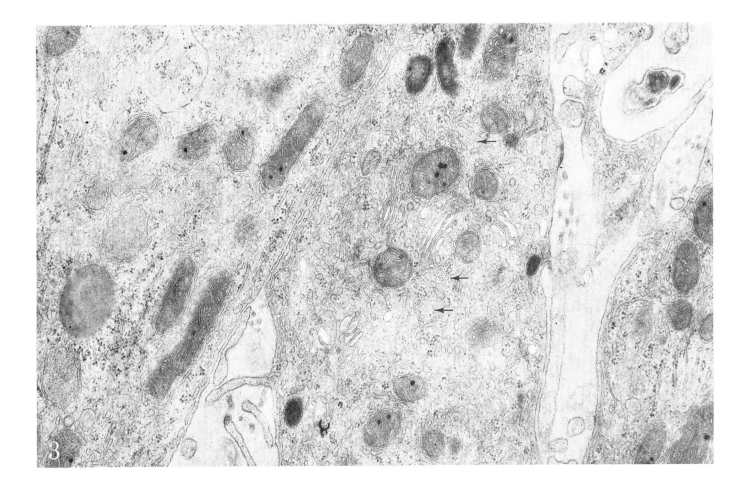

The Testis

1. L. In the rat the seminiferous tubules are supported by a very small amount of interstitial tissue. Groups of interstitial cells of Leydig (ic) and blood vessels occupy triangular spaces between the closely packed seminiferous tubules. These are large, irregular cells with very dense cytoplasm. The nuclei are large, vary in shape from one cell to another, and contain prominent nucleoli. The interstitial cells synthesize the steroid hormone which controls the development of other parts of the reproductive system and of secondary sex characteristics of the male. The seminiferous epithelium of the nearby tubules is made up of supportive cells, the Sertoli or sustentacular cells, and stratified germinal epithelial cells in several of the stages of spermatogenesis. The Sertoli cell (S), located in the basal layer of the epithelium, has a large, pale, indented nucleus and a cytoplasmic cell body which extends to the lumen of the seminiferous tubule. The cytoplasm contains many small dense granules. Examples of the spermatogonia (sg), spermatocytes (sc), and spermatids (st), some of which display a distinct acrosome (ac), can be identified. The spermatids are partially embedded in the apices of Sertoli cells. At the periphery of the lumen are many dense granules or chromatoid bodies. These dense accumulations are in the portion of the spermatid cytoplasm which is discarded when mature spermatozoa are released into the lumen of the seminiferous tubule. 800×

2. From the basal lamina of the epithelium the attenuated cytoplasm of one of the Sertoli cells (S) extends between germinal cells in the early stages of spermatogenesis. A spermatid in a later stage of maturation is embedded in the cytoplasm of the same Sertoli cell. The Sertoli cell cytoplasm contains elongated mitochondria, many small clear vesicles, and small to very large granules. The mitochondria have transversely oriented crests and a dense matrix. The mitochondria of the Sertoli cells contrast remarkably with those of the neighboring germinal cells. As the germinal cells pass through the early stages of spermatogenesis, the mitochondria become progressively more vacuolated. This vacuolation is not present in the Sertoli cell mitochondria. 3250×

The Testis—Interstitial Cell

3. Abundant Golgi complexes are demonstrated in interstitial cells of the testis. Complex tubular systems of agranular endoplasmic reticulum extend through the cytoplasm of both cells. Homogeneous lipid droplets and less dense granules, which contain a finely granular material, are seen in one cell. Smaller dense granules, surrounded by two layers of membrane, are located in the Golgi region and at the periphery of another cell. Many filaments (arrows) have been sectioned transversely and obliquely, particularly at the periphery of the Golgi apparatus. Mitchondria, some with a number of dense bodies, are located within the Golgi zone. 33,000×

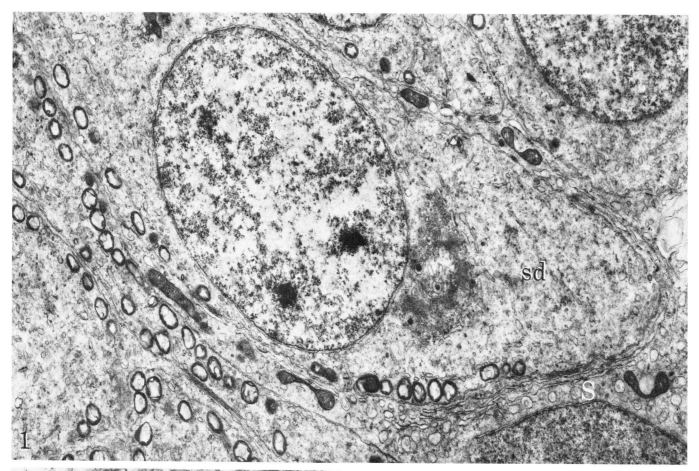

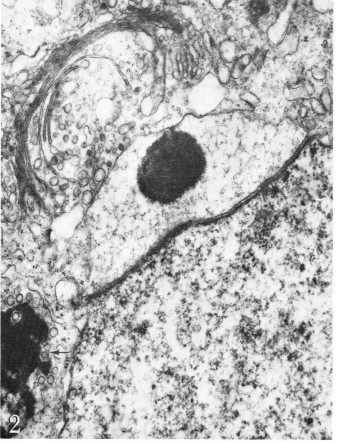

The Testis—The Spermatids

1. A spermatid in the Golgi stage (sd) is embedded in a deep concavity in the nearby Sertoli (S) cell. The elongated or dumbbell-shaped mitochondria indentify the Sertoli cell cytoplasm. The spermatid mitochondria are all vacuolated, the crests having become displaced to the periphery. A considerable amount of moderately dense material occupies the Golgi sacs, and granules are seen in the vesicles. Some dense bodies have developed in the nucleus. 9700×

2. An acrosomal vesicle has developed over the depression in one pole of the spermatid nucleus. A dense layer is seen along the contact zone between them. A number of granules are located between the flattened Golgi sacs and the acrosomal vesicle. A dense granule is found within the acrosomal vesicle. This granule has been reported to have developed by the fusion of the small proacrosomal granules of the Golgi complex. A dense chromatoid body (arrow), bordered by a number of vesicles, has formed near the Golgi complex. 21,000×

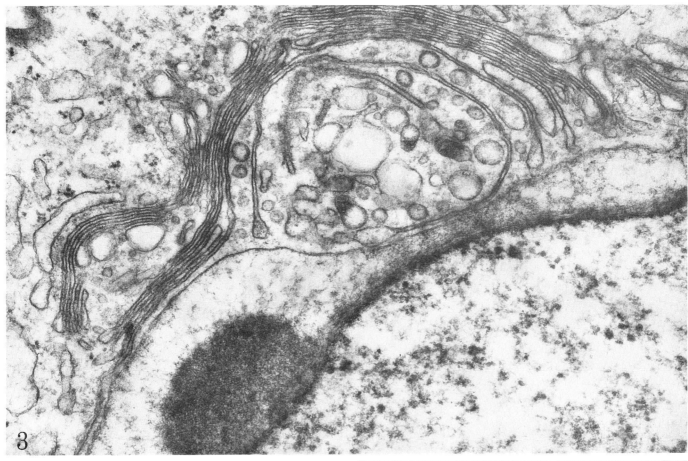

3. The acrosomal granule descends in the vesicle until it becomes attached to the dense material which lines the depression in the pole of the nucleus. The vesicle and the density have spread along the nuclear surface. Filamentous material can be seen in the dense material at the nuclear periphery. The Golgi complex in this stage begins to migrate toward the opposite pole of the nucleus. Granules, vesicles, and sacs appear interconnected in the concavity of the Golgi complex. 50,000 ×

4. In the second or cap phase of spermiogenesis the acrosome becomes flattened against the nuclear surface. The acrosomal vesicle increases its area of contact with the nucleus to form a cape over one pole. The Golgi and the chromatoid body (arrow) have migrated toward the opposite pole of the nucleus, where some large dense granules are seen. The chromatoid body is encircled by a number of very small vesicles. 7500 ×

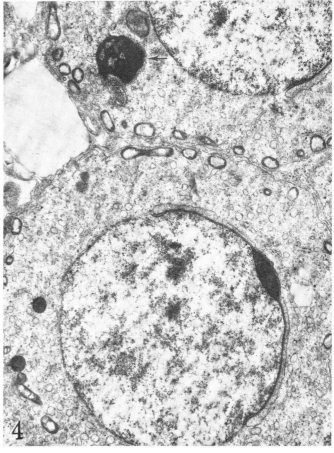

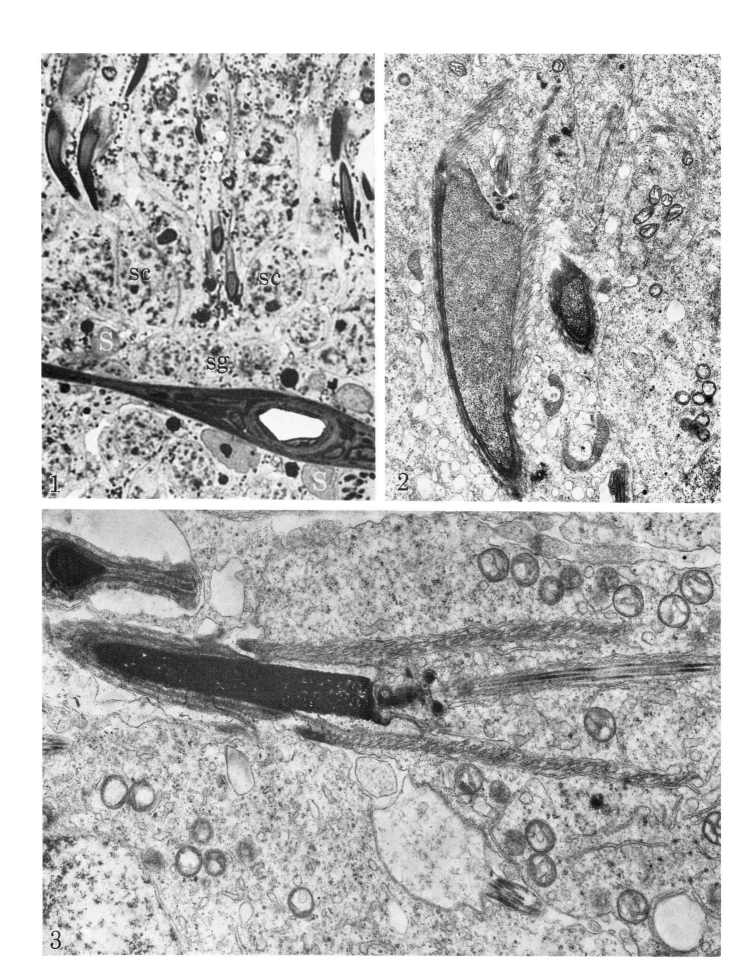

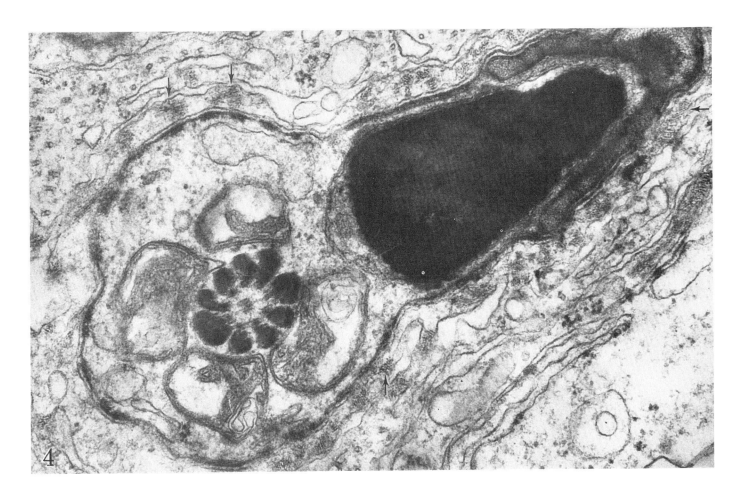

The Testis

1. L. Small-to-large, dense granules are located in the Sertoli (S) cytoplasm. Spermatogonia (sg) in the basal layer and spermatocytes (sc) in the next layer occupy the spaces between the Sertoli cells. Spermatids in the process of maturation are embedded in Sertoli cell cytoplasm. The spermatids, in this phase, have undergone an elongation of the nucleus and the acrosomal cap, and a condensation of the nuclear content. Due to its irregularity of shape, the caudal end of the nucleus appears oval in some planes of section and square in others. In some spermatids two moderately dense lines extend out from the caudal border of the nucleus. 1000×

2. The acrosomal cap has begun to develop a distinct prolongation and envelops the two aspects of the nucleus quite unequally. The caudal sheath, composed of a large number of cytoplasmic microtubules, extends out from the border of the cap. Two dense concavities are seen at the caudal surface of the nucleus. The centrioles have migrated to this vicinity before this time. The distal centriole, oriented perpendicularly to the nuclear surface, has developed a flagellum. 7500×

3. The apex of the acrosome becomes extremely elongated, and the network within the nucleoplasm stains considerably more intensely than the acrosomal cap. With the continued maturation of the spermatid the condensation of the nuclear content progresses by a thickening of the filamentous material at the expense of the lighter background. Dense fibers have developed in what will become the middle piece of the tail of a spermatozoon. During these advanced phases of the maturation of the spermatid, increasingly large vesicular spaces appear around the apex of the nucleus. 16,250×

4. A section has passed obliquely through the nucleus and almost transversely through the flagellum. In this middle piece of the developing spermatozoon the two central microtubules or fibrils are surrounded by the nine doublets typical of cilia or flagella. These, together, are termed the axial filament complex. The axial complex is, in turn, surrounded by nine large, dense bodies which are the outer dense longitudinally oriented fibers as seen in transverse section. A mitochondrial sheath surrounds this sleeve of dense fibers. A number of profiles of the agranular endoplasmic reticulum are found between the mitochondrial sheath and the membrane, which now distinctly outlines the future spermatozoon. The membrane of the spermatozoal portion of the spermatid is extremely dense, while that lining the residual portion is considerably less prominent. This interspace is the plane of cleavage along which the spermatozoon separates from the residual cytoplasm. A large number of the filaments (arrows), which were seen in an accessory cap covering the acrosomal vesicle, are still to be found within the first layer of the residual cytoplasm surrounding the developing spermatozoon. 65,000×

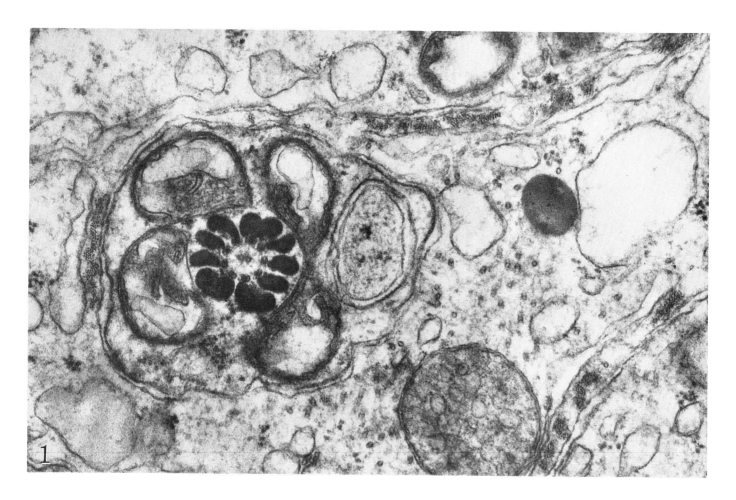

The Testis

1. A section has passed through the middle piece of the future spermatozoon. The mitochondrial sheath encircles the nine outer dense fibers. Beyond the mitochondrial sheath is a double layer of membranes which delineates the plane of cleavage of this part of the tail from the residual cytoplasm. Transversely sectioned filaments in an interrupted layer extend out into the latter. Numerous, transversely sectioned cytoplasmic microtubules, large vesicles, a multivesicular body with two surrounding membranes, and a dense granule are located in the residual cytoplasm. 70,000×

2. A section through the lumen of a seminiferous tubule contains segments of middle (m), principal (p), and end (e) pieces of sperm tail in oblique and transverse section. 4500×

3. In frontal view, some cytoplasm still surrounds the acrosome and nucleus of a spermatozoon. The middle pieces of the spermatozoa consist of central fibrils (cf), the peripheral fibrils (pf), the outer dense fibers (of), a mitochondrial sheath (ms), a cytoplasmic layer, and a plasma membrane. The mitochondria are arranged in a closely packed circle around the sheath composed of the nine outer dense fibers. 18,000×

4. In the principal piece of the sperm tail a dense fibrous sheath has replaced the mitochondrial sheath. Two of the outer dense fibers appear to have become incorporated into equatorial expansions, the longitudinal columns, of this fiber sheath. The columns divide the principal piece into a major and a minor compartment, with three remaining dense fibers included in the minor and four within the major compartment. The two longitudinal columns extend in continuity through most of the principal piece. They are interconnected by semicircular fibrils which complete the fibrous sheath. A few end pieces (e) of the tail, wherein the outer dense fibers and the fibrous sheath are no longer present, have been sectioned slightly obliquely. 65,000×

5. In a transverse section of each of several end pieces of sperm tail two typical central microtubules and the nine doublets (outer fibrils) can be visualized. Fibrillar radiations pass from the central microtubules toward each of the doublets. In the proximal portion of the end pieces some evidence of the two longitudinal columns of the fibrous sheath still persist. 60,000×

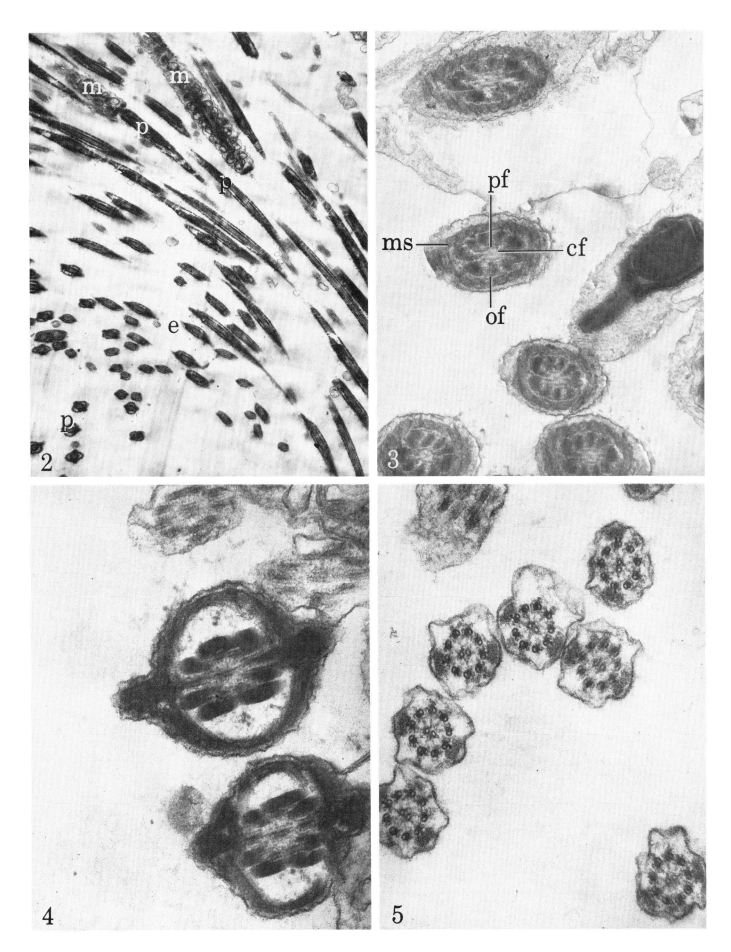

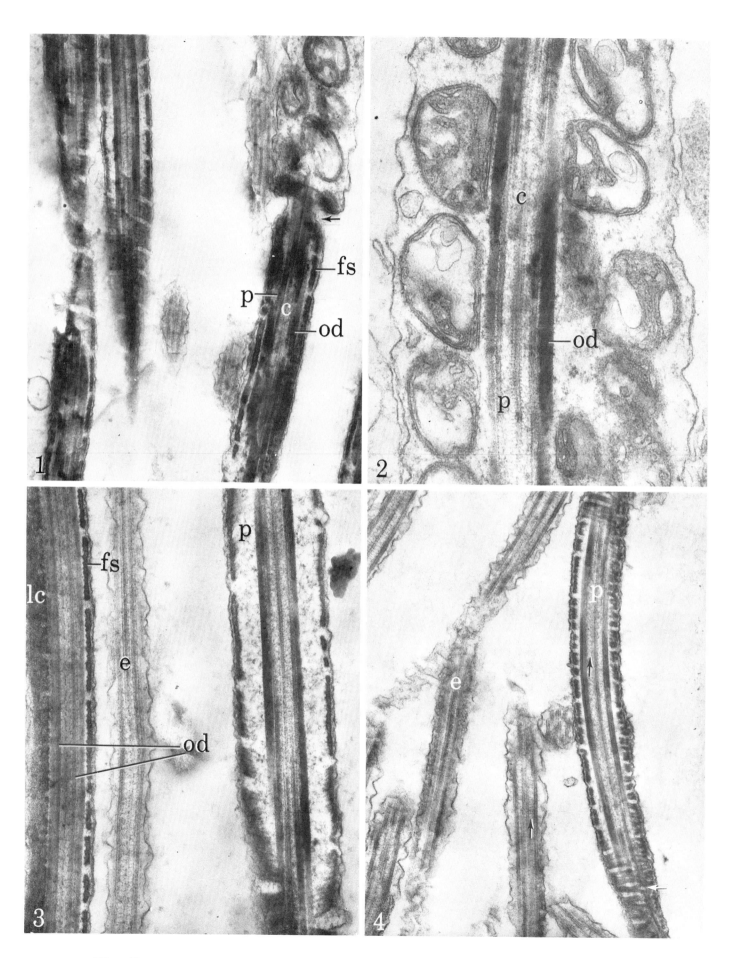

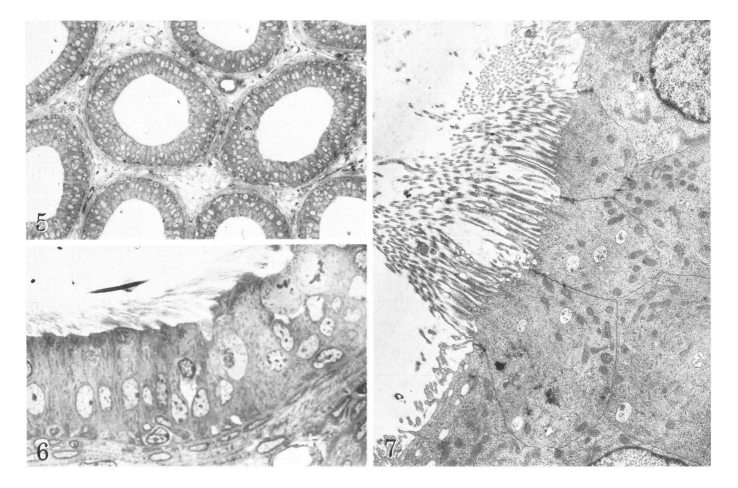

The Testis—The Spermatozoa

1. In a longitudinal section the transformation from the caudal portion of the middle piece to the principal piece can be seen (arrow). The mitochondria do not continue into the principal piece, but the central microtubules (c), the peripheral doublets (p), and the outer dense fibers (od) are continuous from the middle to the principal portion of the tail. In the principal piece a new, interrupted layer of dense material, which has been termed the fibrous sheath (fs), replaces the mitochondrial sheath beneath the surrounding plasma membrane. Other obliquely sectioned parts of principal and end pieces from other spermatozoa are seen. 25,000×

2. The middle piece, in longitudinal section, shows the mitochondria aligned between the irregular plasma membrane and the outer dense (od) fibers. The central microtubules (c) and the peripheral doublets (p) of the sperm tail have a striated surface. 60,000×

3. An end piece (e) is located between the principal pieces (p) of two other spermatozoa. Neither the outer dense fiber nor the peripheral fibrous sheath extends into the end piece. In one of the principal pieces the section includes the interrupted fibrous sheath (fs), several of the outer dense fibers (od), and a wide, dense band which is a longitudinal column (lc). In the other principal piece some of the riblike interconnections between the longitudinal columns can be seen in surface view. These have been described as hoops which encircle this segment of the spermatozoon. 42,500×

4. An end piece has been sectioned almost transversely between longitudinally sectioned end (e) and principal (p) pieces of

spermatozoon. Filamentous bridges (arrows) can be seen between the central microtubules and the peripheral doublets in the end pieces, and, less prominently, in the principal piece. At the two ends of the principal piece, surface views of the riblike hoops of the fibrous sheath have been obtained. Some of these ribs are disposed slightly obliquely to the transverse axis of the tail and one appears to branch (white arrow). 34,000×

The Ductus Epididymus

5. L. The ductus epididymus is an extremely tortuous, circular tube lined by columnar epithelium. There is an interrupted layer of basal cells with pale cytoplasm between the epithelium and the basal lamina. The numerous capillaries in the loose connective tissue have been emptied by perfusion. 150×

6. L. The columnar epithelial cells have large, pale nuclei. This epithelium has long, thin microvilli on its luminal surface. These were termed stereocilia from their appearance in the light microscope. The cell in early telophase demonstrates the typical elevation of its base during mitosis. Thus cell division occurs at the border of the lumen. There is no evidence of microvilli on the cell in mitosis. The small basal cells have a pale cytoplasm. A part of the head of a spermatozoon is seen in longitudinal section in the lumen, and a few tail pieces have been sectioned transversely. 680×

7. The long microvilli arise from the central, major part of the apex of each cell. The microvilli at the borders are shorter. The apices of the cells contain a number of moderately large vesicles with irregularly stained content and a few dense granules. 6000×

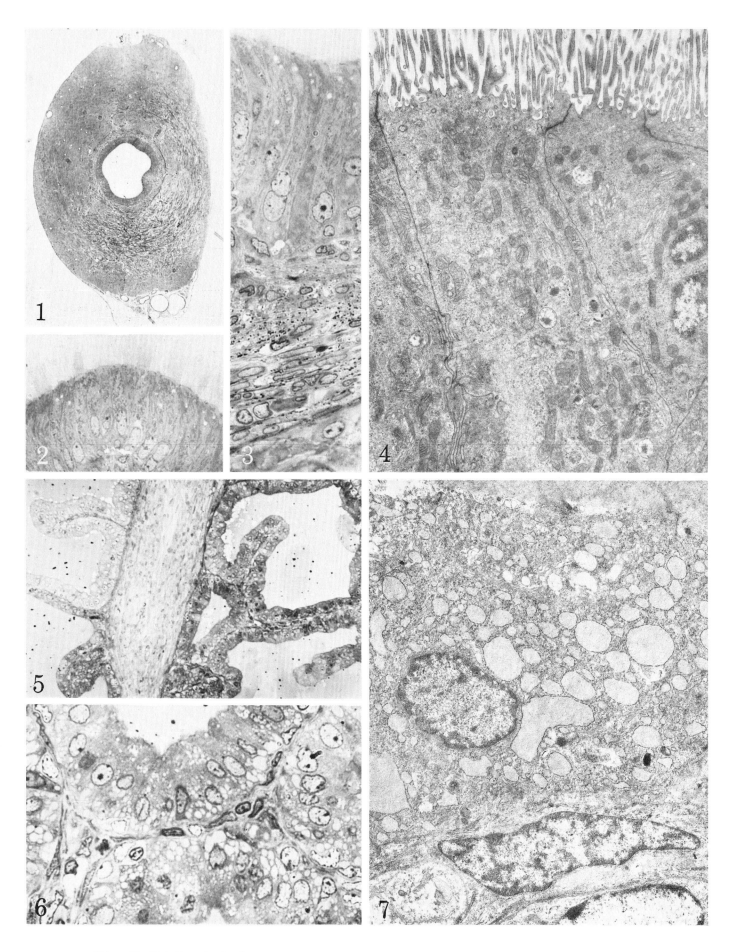

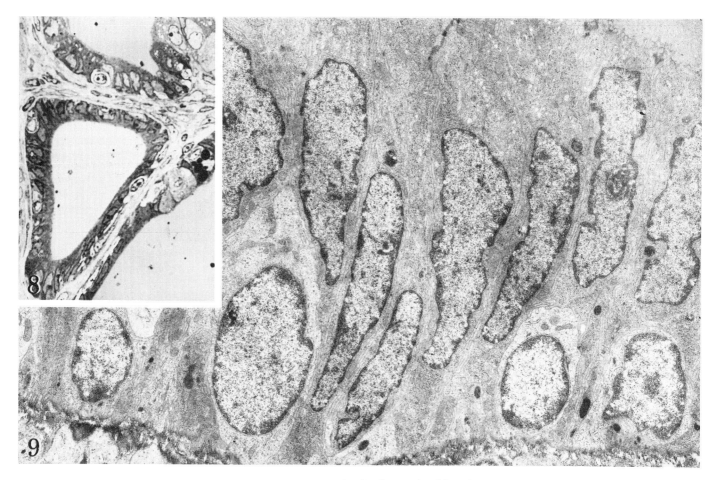

The Ductus Deferens

1. L. The ductus (vas) deferens is a tube with a mucosal lining, a thick muscularis, and an adventitia. 40 ×

2. L. In the initial part of the duct, the epithelium is crowned by very tall, thin microvilli similar to those seen in the epididymus. Mitochondria, vacuoles, and an occasional dense granule can be seen in the cytoplasm. 560 ×

3. L. At a greater distance from the epididymus, long microvilli are no longer seen on a tall pseudostratified epithelium. The epithelium is supported by a thick band of connective tissue, the lamina propria. This consists of fibrous tissue with a high content of densely stained elastic fibers. The elastic fibers are particularly abundant in a central band in the lamina propria. 640 ×

4. The supranuclear portion of the cells contains many moderately dense mitochondria, an abundant Golgi apparatus which often extends well into the apex of the cell, and vacuoles with moderate-to-dense content. The mitochondrial membranes, including the crests, are accentuated by a dense material. 6000 ×

The Seminal Vesicles

5. L. Each seminal vesicle is a coiled tube, the wall of which contains smooth muscle and connective tissue circumferential layers. The extremely folded epithelial border appears as a labyrinth of secretion-filled chambers. The secretion is a viscous solution high in globulin and fructose content. A considerable level of the hormone prostaglandin has been found in this secretory material. 150 ×

6. L. The epithelial cell cytoplasm, which differs markedly in density, is vacuolated from base to apex. There is a thin layer of connective tissue between the epithelial layers. 680 ×

7. The vacuolated appearance of the cytoplasm is due to huge distended cisterns of the granular endoplasmic reticulum. There is an extensive clear Golgi apparatus below the cell apex. Some dense granules are seen near the base. A smooth muscle fiber is seen at the base of the epithelium. 5400 ×

The Prostate

The prostate consists of many tubuloalveolar glands in a fibromuscular stroma. The secretion contains citric acid and several enzymes, including diastase, β-glucuronidase, acid phosphatase, other proteolytic enzymes, and fibrinolysin. The secretion is emptied into the prostatic urethra by a number of ducts which have numerous tributaries. A hormone, prostaglandin, has been isolated from the seminal fluid and was originally considered specific for the prostate. Very recently this was demonstrated to be one of a family of vasopressor hormones elaborated by a number of organs throughout the body.

8. L. There is a wide variation in form, size, and density in the epithelial cells which line these glands. They range from cuboidal to pseudostratified columnar and from small, dense, to large, pale cells with a vesicular cytoplasm. There are many very pale basal cells. Other very large, pale cells reach the lumen of the gland. Densities appear against a lighter background in the lumen. 640 ×

9. In the epithelium, the cells are moderate in height with elongated irregular nuclei, abundant granular endoplasmic reticulum, and a number of dense granules throughout the cytoplasm. The lighter basal cells, with large nuclei located between the bases of the columnar cells, are numerous. 5600 ×

The Male Reproductive System   155

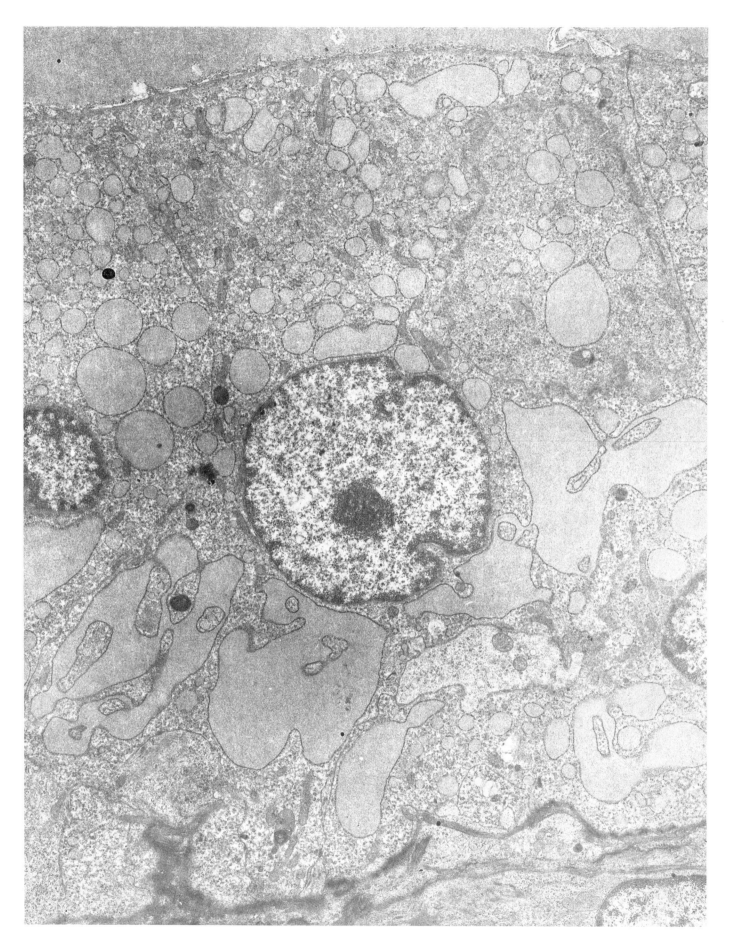

The Prostate

The actively secreting, prostatic epithelial cells contain an interconnected network of cisternae of the granular endoplasmic reticulum. A number of dense granules, the majority in the juxtanuclear zone, are seen in the cytoplasm. Short apical microvilli are embedded in a moderately dense luminal content. Elastic fibers are seen among the underlying fibroblasts and smooth muscle cells. 16,000 ×

REFERENCES

Bawa, S. R. (1963). The fine structure of the sertoli cells of the human testis. *J. Ultrastruct. Res.* **9**, 459–474.

Belt, W. D., and Cavazos, L. F. (1967). Fine structure of the interstitial cells of Leydig in the boar. *Anat. Record* **158**, 333–350.

Brandes, D. (1966). The fine structure and histochemistry of prostatic glands in relation to sex hormones. *Intern. Rev. Cytol.* **20**, 207–276.

Bröckelman, J. (1963). Fine structure of germ cells and Sertoli cells during the cycle of the seminiferous epithelium in the rat. *Z. Zellforsch. Mikroskop. Anat.* **59**, 820–850.

Burgos, M. H., and Fawcett, D. W. (1956). An electron microscope study of spermatid differentiation in the toad, Bufo arenarum Hensel. *J. Biophys. Biochem. Cytol.* **2**, 223–240.

Burgos, M. H., and Vitale-Calpe, R. (1967). The fine structure of the Sertoli cell-spermatozoan relationship in the toad. *J. Ultrastruct. Res.* **19**, 221–237.

Christensen, A. K. (1965). Fine structure of testicular interstitial cells in the guinea pig. *J. Cell Biol.* **26**, 911–935.

Christensen, A. K., and Fawcett, D. W. (1966). The fine structure of the interstitial cells of the mouse testis. *Am. J. Anat.* **118**, 551–571.

Crabo, B. (1963). Fine structure of the interstitial cells of the rabbit testis. *Z. Zellforsch. Mikroskop. Anat.* **61**, 587–604.

Fawcett, D. W. (1965). The anatomy of the mammalian spermatozoan with particular reference to the guinea pig. *Z. Zellforsch. Mikroskop. Anat.* **67**, 279–296.

Fawcett, D. W., and Ito, S. (1965). The fine structure of bat spermatozoa. *Am. J. Anat.* **116**, 567–610.

Frank, A. L., and Christensen, A. K. (1968). Localization of acid phosphatase in lipofuschin granules and possible autophagic vacuoles in interstitial cells of the guinea pig testis. *J. Cell Biol.* **36**, 1–13.

Friend, D. S., and Farquhar, M. G. (1967). Functions of coated vesicles during protein absorption in the rat vas deferens. *J. Cell Biol.* **35**, 357–376.

Horstmann, E., Richer, R., and Roosen-Runge, E. (1966). Zur elektronen mikroskopie der Kernein schlüsse im menschlichen Nebenhodenepithel. *Z. Zellforsch. Mikroskop. Anat.* **69**, 69–79.

Krishan, A., and Buck, R. C. (1965). Ultrastructure of cell division in insect spermatogenesis. *J. Ultrastruct. Res.* **13**, 444–458.

Leeson, T. S., and Leeson, C. R. (1965). The fine structure of cavernous tissue in the adult rat penis. *Invest. Urol.* **3**, 144–158.

McLaren, A., ed. (1966–1967). "Advances in Reproductive Physiology," Vols. 1 and 2. Academic Press, New York.

Makita, T., and Kiwaki, S. (1966). The fine structure of the muscle cell of the mouse vas deferens. *Bull. Fac. Agr., Yamaguchi Univ.* **17**, 853–872.

Mao, P., and Nakao, K. (1966). Variation of localization of AMPase and ATPase activities at the plasma membrane of human prostatic epithelial cells: An electron microscopic study. *J. Histochem. Cytochem.* **14**, 203–204.

Nagano, T. (1965). Localization of adenosine triphosphatase activity in the rat sperm tail as revealed by electron microscopy. *J. Cell Biol.* **25**, 101–112.

Nagano, T. (1967). Male genital organ. *In* "Fine Structure of Cells and Tissues Electron Miscrosopic Atlas" (H. Sakaguchi et al., eds.), Vol. 3, pp. 63–120. Igaku Shoin, Tokyo.

Ortavant, R. (1958). Le cycle spermatogénétique chez le Bélier. Thèse, Fac. Sci., Université de Paris.

Roosen-Runge, E. C., and Giesel, L. O. (1950). Quantitative studies on spermatogenesis in the albino rat. *Am, J. Anat.* **87**, 1–30.

Schmidt, F. C. (1964). Licht-und elektronemikroskopische Untersuchungen am menschlichen Hoden und Nebenhoden. *Z. Zellforsch. Mikroskop. Anat.* **63**, 707–727.

Tandler, B., and Moriber, L. G. (1966). Microtubular structures associated with the acrosome during spermiogenesis in the water-strider, Gerris remigis (Say). *J. Ultrastruct. Res.* **14**, 391–404.

Threadgold, L. T. (1967). *In* "The Ultrastructure of the Animal Cell," Intern. Ser. Monographs Pure Appl. Biol. Zool., Div., Vol. 37, pp. 296–306. Pergamon Press, Oxford.

THE FEMALE
REPRODUCTIVE
SYSTEM

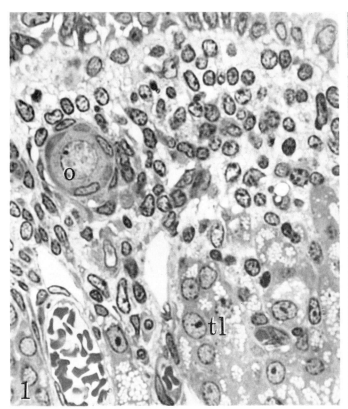

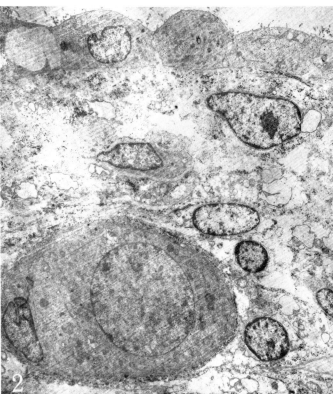

The Ovary

1. L. The ovary has a continuous surface layer of cuboidal cells which has been termed the germinal epithelium. The term persists from the belief that primordial oocytes originated from this layer. Deep to the epithelium is a layer of connective tissue, the tunica albuginea. Within the stroma deep to the tunica albuginea the ovarian follicles develop from primordial into primary follicles. In the primordial follicle the large, pale oocyte (o) is surrounded by a single layer of flattened follicular cells (f). At a deeper level collections of moderately large cells with numerous pale cytoplasmic vacuoles are part of a corpus luteum. A few small dense granules can be seen in the theca luteal cells (tl). Several corpora lutea develop simultaneously in the rat. 720×

2. The epithelial cells of the ovary bulge into the peritoneal cavity. A small number of microvilli project from the free surface. The collagen in the tunica albuginea is in loosely arranged bundles. A primordial follicle within the superficial stroma of the ovary is surrounded by a thin layer of collagen fibrils. The thin layer of cytoplasm of the follicular cells which surrounds the oocyte is difficult to distinguish since, at this stage, the two types of cells have approximately the same cytoplasmic density. 2500×

3. L. The secondary follicle becomes oval in shape. A layer of amorphous neutral protein–polysaccharide has been deposited about the periphery of the oocyte. This is termed the zona pellucida (zp). The cytoplasm of the oocyte is abundant, and the chromatin appears as dense masses in the very pale nucleus. A thick layer of granulosa cells surrounds the zona pellucida. An extremely high number of cells in mitosis are seen in this layer. A cavity, the antrum, has developed in the stratum granulosum of the larger follicle. The content of the antrum is called liquor folliculi. 180×

4. L. The zona pellucida forms a wide, homogeneous band around the cytoplasm of the oocyte. In the stratum granulosum, cells can be seen in various stages of mitosis. 720×

5. Considerable spaces intervene between the cells of the granulosa, one of which is in anaphase. It is a confluence of these spaces which creates the antrum. The zona pellucida is interlaced by moderately large cytoplasmic processes which extend from the granulosa cells (arrows). Numerous short microvilli protrude from the surface of the oocyte. 2500×

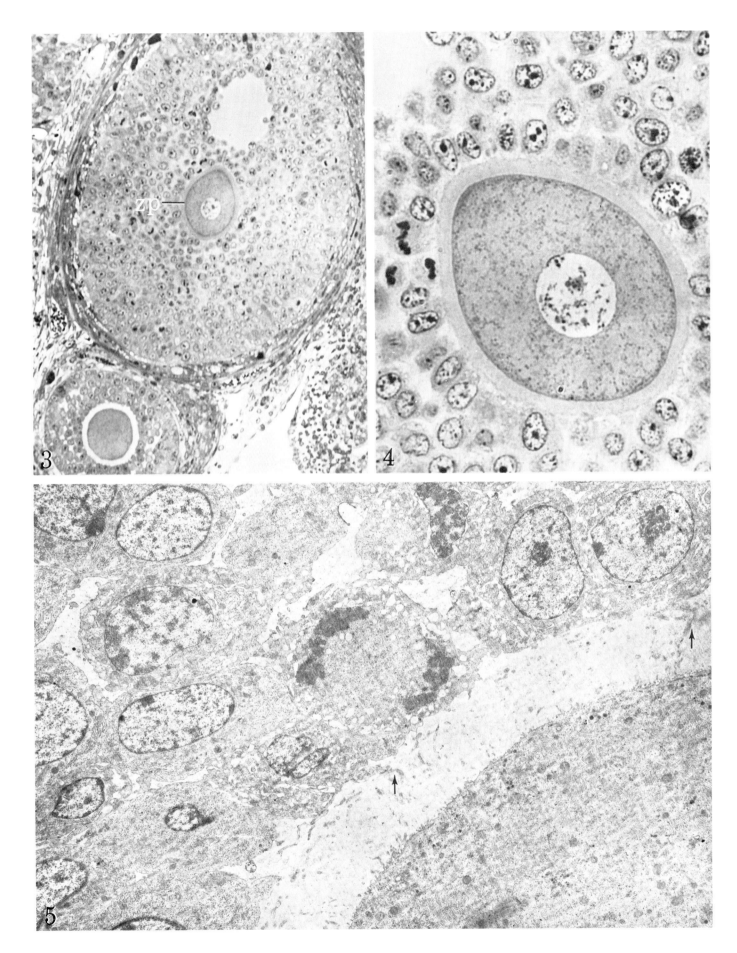

3

4

zp

5

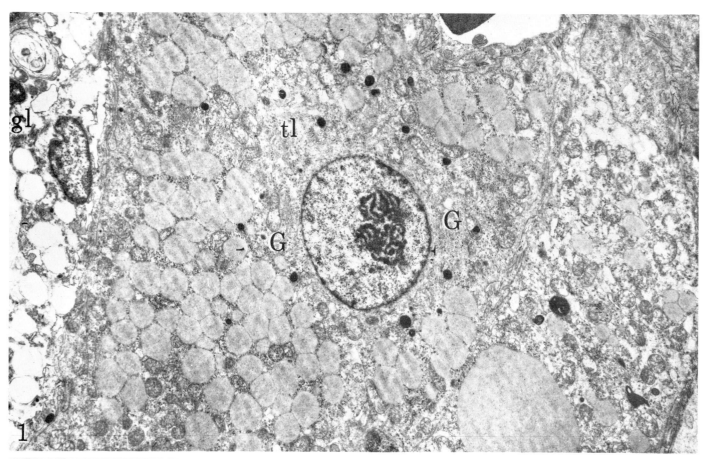

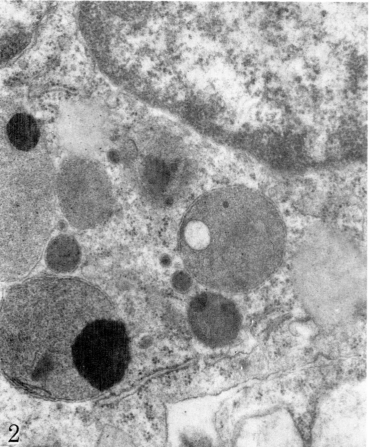

The Ovary—Corpus Luteum

1. After ovulation, the follicular cells develop into granulosa luteal cells (gl). These are cells which appear remarkably vesiculated since their content is not retained during fixation. Outside the basal lamina of the follicle another type of cell, the theca luteal cell, develops from the theca interna layer. The theca luteal cells (tl) contain many large, lightly stained granules and a number of smaller dense granules. They have a rounded nucleus with a very large nucleolus. These cells have multiple Golgi (G) complexes. A large dense granule is located in a mitochondrion. 7800×

The Ovary—Theca Luteal Cells

2. Composite types of granules can be seen in some theca luteal cells. Some granules have a pale homogeneous background similar to the lipid droplets, while in others the matrix is either finely granular or quite variable with localized, internal membrane-like profiles (arrow). 32,000×

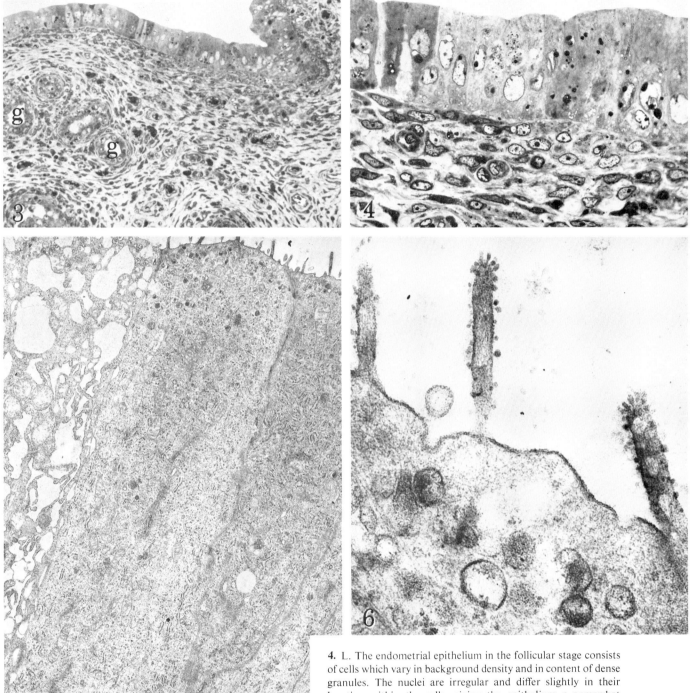

The Uterus

3. L. The lining of the uterus, commonly known as endometrium, is a simple columnar epithelium which is invaginated to form numerous deep tubular glands. These glands (g) extend into the lamina propria and, when sectioned transversely, appear circular. The lining cells of the glands are not as tall as those on the surface epithelium. The endometrial stroma is highly cellular with a rich supply of blood vessels. 140×

4. L. The endometrial epithelium in the follicular stage consists of cells which vary in background density and in content of dense granules. The nuclei are irregular and differ slightly in their location within the cells, giving the epithelium a somewhat pseudostratified appearance. A number of vacuoles are seen in some of the cells. 560×

5. In general the tall columnar epithelial cells have a high content of granular endoplasmic reticulum and extensive Golgi zones. In the apex of the darker cells small vesicles are wholly or partly filled with moderately dense content. The vacuolated cells appear to contain a network of interconnected channels with very pale content. A few small dense granules are found in the cytoplasm. 5600×

6. Globular densities are found on the surfaces of a number of the apical microvilli on these cells in the uterine epithelium. Some of the apical vesicles are elongated or flask-shaped and their content varies in density. Filaments can be seen between these vesicles and the apical surface of the cell. 64,000×

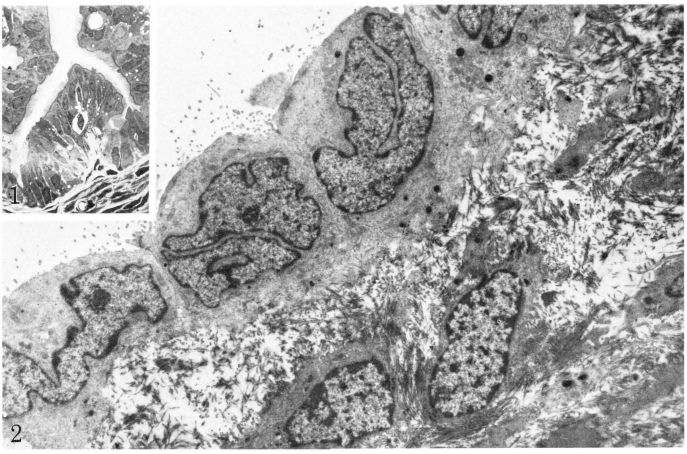

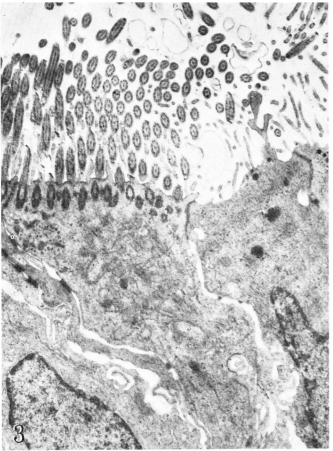

The Oviduct (Uterine Tube)

1. L. In the distal portion of the oviduct the deep folds are lined by a columnar epithelium. In this portion of the tube most of the cells are ciliated. 520×

2. Some epithelial cells of the oviduct near the uterus are cuboidal in shape with very large, irregular nuclei. All of these cells have apical microvilli. Many large fibroblasts are found in the subepithelial loose connective tissue. 9100×

3. Some epithelial cells are ciliated while others bear only the long thin microvilli. A group of dense granules and of smaller glycogen granules are found in the cell apex beneath the basal bodies of the cilia. 11,350×

The Mammary Gland

4. L. In the lactating mammary gland, fixed by perfusion of the whole animal, small-to-large lipid droplets (l) and vacuoles containing small dense granules occupy much of the cytoplasm of many of the cells. Fine dense granules and larger, lighter, homogeneous fat droplets can be seen within the lumen. Myoepithelial cells (arrows) are stretched out along the bases of the alveolar epithelial cells. 500×

5. The epithelial cell of the lactating gland has a moderately regular, small nucleus. The cytoplasm contains many layers of granular endoplasmic reticulum, prominent Golgi complexes, large secretory vacuoles, and lipid droplets (l). The vacuoles contain very fine granules in the Golgi zone. These granules, which increase in density and size as the vacuoles approach the

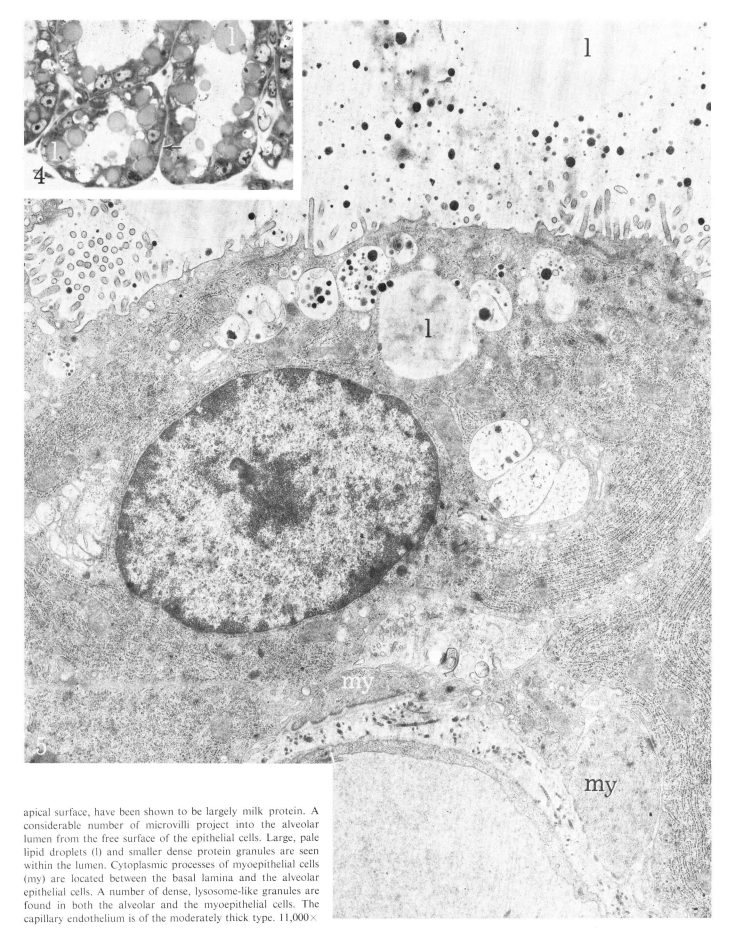

apical surface, have been shown to be largely milk protein. A considerable number of microvilli project into the alveolar lumen from the free surface of the epithelial cells. Large, pale lipid droplets (l) and smaller dense protein granules are seen within the lumen. Cytoplasmic processes of myoepithelial cells (my) are located between the basal lamina and the alveolar epithelial cells. A number of dense, lysosome-like granules are found in both the alveolar and the myoepithelial cells. The capillary endothelium is of the moderately thick type. 11,000×

REFERENCES

OVARY, OVIDUCT, AND UTERUS

Adams, E. C., and Hertig, A. T. (1964). Studies on guinea pig oocytes. I. Electron microscopic observations on the development of cytoplasmic organelles in oocytes of primordial and primary follicles. *J. Cell Biol.* **21**, 397–427.

Baker, T. G., and Franchi, L. L. (1967). The fine structure of oogonia and oocytes in human ovaries. *J. Cell Sci.* **2**, 213–224.

Bjorkman, N., and Fredricsson, B. (1961). The bovine oviduct epithelium and its secretory process as studied with the electron microscope and histochemical tests. *Z. Zellforsch. Mikroskop. Anat.* **55**, 500–513.

Blanchette, E. J. (1966). Ovarian steroid cells. II. The lutein cell. *J. Cell. Biol.* **31**, 517–542.

Clyman, M. J. (1963). Electron microscopy of the human endometrium. *Progr. Gynecol.* **4**, 36–57.

Clyman, M. J. (1966). Electron microscopy of the human fallopian tube. *Fertility, Sterility* **17**, 281–301.

Enders, A. C., and Lyons, R. W. (1964). Observations on the fine structure of lutein cells. II. The effects of hypophysectomy and mammotrophic hormones in the rat. *J. Cell Biol.* **22**, 127–141.

Flaks, B., and Bresloff, P. (1966). Some observations on the fine structure of the lutein cells of X-irradiated rat ovary. *J. Cell Biol.* **30**, 227–236.

Green, J. A., Garcilazo, J. A., and Maqueo, M. (1968). Ultrastructure of the human ovary. *Am. J. Obstet. Gynecol.* **102**, 57–64.

Greep, R. O. (1963). Histology, histochemistry and ultrastructure of adult ovary. *In* "The Ovary" (H. G. Grady, and D. E. Smith, eds.), pp. 48–68. Williams & Wilkins, Baltimore, Maryland.

Nilsson, O. (1962). Electron microscopy of the glandular epithelium in the human uterus. I. Follicular phase. *J. Ultrastruct. Res.* **6**, 413–421.

Nilsson, O. (1962). Electron microscopy of the glandular epithelium in the human uterus. II. Early and late luteal phase. *J. Ultrastruct. Res.* **6**, 422–431.

Price, D., and Williams-Ashman, H. G. (1961). Accessory mammalian reproductive glands. *In* "Sex and Internal Secretions" (W. C. Young, and G. W. Corner, eds.), 3rd ed., vol. 1, pp. 366–448. Wilkins, Baltimore, Maryland.

Wischnitzer, S. (1965). The ultrastructure of the germinal epithelium of the mouse ovary. *J. Morphol.* **117**, 387–399.

Zamboni, L., and Gondos, B. (1968). Intercellular bridges and synchronization of germ cell differentiation during oogenesis in the rabbit. *J. Cell. Biol.* **36**, 276–282.

Zuckerman, S., ed. (1962). "The Ovary," Vols. 1 and 2. Academic Press, New York.

MAMMARY GLAND

Cowie, A. T., and Folley, S. J. (1961). The mammary gland and lactation. *In* "Sex and Internal Secretions" (W. C. Young, and G. W. Corner, eds.), 3rd ed., vol. 1, p. 590. Williams & Wilkins, Baltimore, Maryland.

Helminen, H. J., and Ericcson, J. L. E. (1968). Studies on mammary gland involution. I. On the ultrastructure of the lactating mammary gland. *J. Ultrastruct. Res.* **25**, 193–213.

Hollmann, K. H. (1959). L'ultrastructure de la glande mammaire normale de la souris en lactation. Etude au microscope electronique. *J. Ultrastruct. Res.* **2**, 423–443.

Hollmann, K. H. (1966). Sur des aspects particuliers des proteines élaborées dans la glande mammaire. Etude au microscope électronique chez la lapine en lactation. *Z. Zellforsch. Mikroskops. Anat.* **69**, 395–402.

Mayer, G. (1961). Histology and cytology of the mammary gland. *In* "Milk: The Mammary Gland and its Secretion" (S. K. Kon, and A. T. Cowie, eds.), Vol. 1, pp. 47–126. Academic Press, New York.

Wellings, S. R., DeOme, K. B., and Pitelka, D. R. (1960). Electron microscopy of milk secretion in the mammary gland of the C^3H/Cngl mouse. I. Cytomorphology of the prelactating and the lactating gland. *J, Natl. Cancer Inst.* **25**, 393–422.

Wellings, S. R., Grunbaum, B. W., and DeOme, K. B. (1960). Electron microscopy of milk secretion in the mammary gland of the C^3H/Cngl mouse. II. Identification of fat and protein particles in milk and in tissue. *J. Natl. Cancer Inst.* **25**, 423–438.

Wellings, S. R., and Phelp, J. R. (1964). The function of the Golgi apparatus in lactating cells of the BALB/cCngl mouse: An electron microscopic and autoradiographic study. *Z. Zellforsch. Mikroskop. Anat.* **61**, 871–882.

THE ENDOCRINE
GLANDS

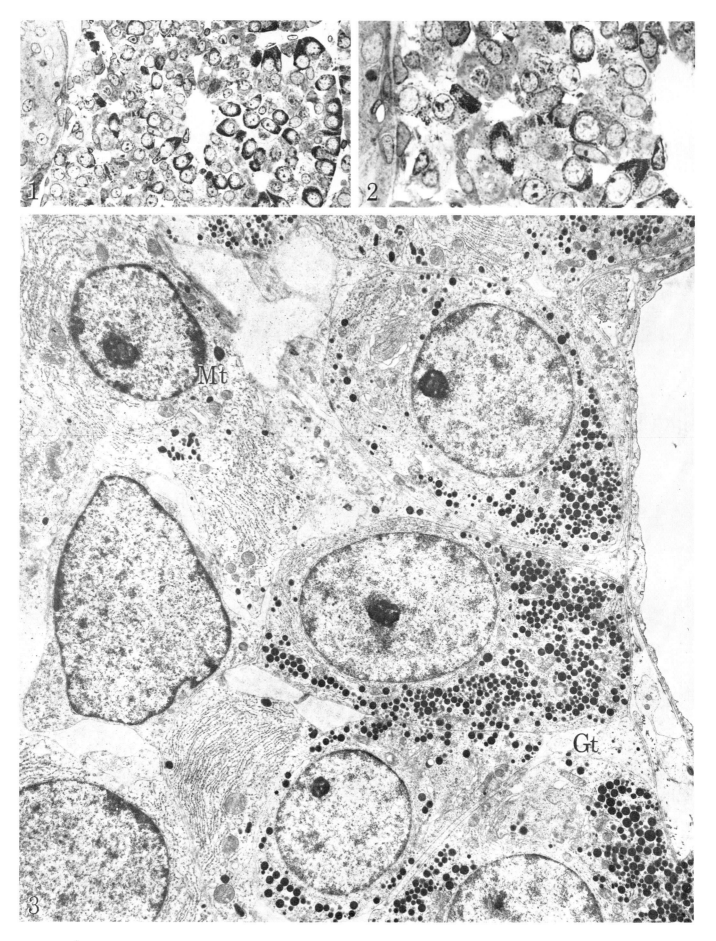

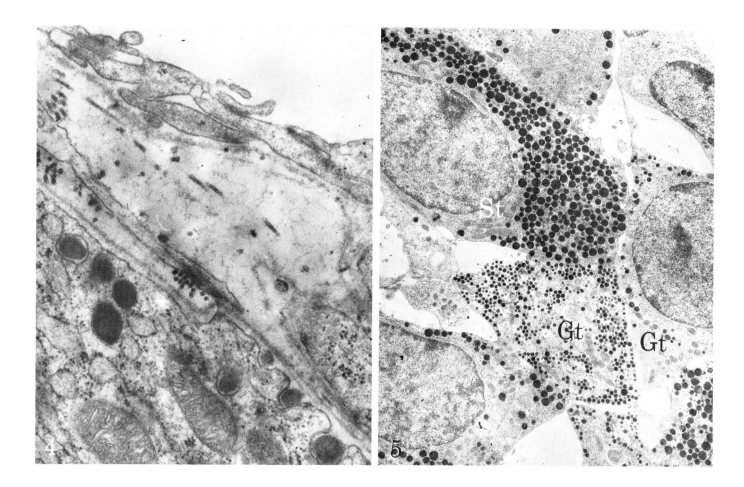

The Hypopthysis—Pars Distalis

1. L. In the pars distalis there is a great variation in the size, shape, and granule content of the various cells. The cells, which are moderately loosely arranged, are interlaced by a rich vascular network. Facing the network of capillaries are a number of cells of extreme density. This appearance is produced by a very large number of closely packed, dense granules. Cells more distant from the capillaries appear less granulated. 300×

2. L. At higher magnification, variations in the number, size, and distribution of granules within cells of the pars distalis can be observed. 600×

At least six hormones are known to be produced by the pars distalis. Most of the specific cells responsible for the synthesis of these hormones have been identified from the changes in their ultrastructural appearance after experimental procedures. The different cells can be identified by the size, shape, and content of their granules.

3. Elongated or ovoid cells, with numerous circular dense granules from 0.15–0.5 μ in diameter, which border on the perivascular space are somatotropes which elaborate the growth stimulating hormone. These cells have large, rounded nuclei with finely granular chromatin and prominent nucleoli. In one of these somatotropes the granular endoplasmic reticulum is layered in a localized juxtanuclear zone. In others the reticulum appears more widely distributed. Another cell, which contains elongated granules from 0.1–0.2 μ in width by 0.2–0.6 μ in length, elaborates mammotropic or lactotropic hormones. There has been considerable discussion regarding the terminol-

ogy for this cell since it has been found to produce luteotropic hormone in some animals such as the rat. The terms lactotrope or luteotrope have both been employed. However, it would appear that it is a mammotrope cell (Mt) in most species of animals. The endoplasmic reticulum, which shows a moderate dilatation of its cisternae, is distributed widely throughout the cytoplasm. The pale cytoplasm with much smaller, dense granules is part of a gonadotrope cell (Gt). The endothelium of the capillaries in the pars distalis is of the fenestrated type. 5200×

4. At the vascular pole of one of these somatotrope cells a moderately dense content can be found in indentations in the cell surface. This can be interpreted with a reasonable degree of certainty as secretory material being released into the extracellular space. The granular form and density are lost by dispersion in the perivascular interstitial fluid. 32,000×

5. Gonadotropes (Gt) are cells with rounded granules which are smaller than those of the somatotrope (St). The majority of the granules in the gonadotropes are less than 0.2 μ in diameter. Two types of gonadotrope cells, those which secrete follicle stimulating hormone (FSH) and those which produce luteinizing hormone (LH) have been described. The former are reported to have a more vesiculated cytoplasm and a number of granules as large as 1 μ. It has also been suggested that the same cell may be responsible for both hormones. In the normal rat, fixed by perfusion with aldehydes, the difference is not remarkable. In the male, the gonadotrope secretes the interstitial cell stimulating hormone (ICSH). 6900×

1. A very irregularly shaped cell with a similar conformity of its nucleus enfolds a cytoplasmic process of another cell. The nucleus contains very dense, granular chromatin and a prominent nucleolus. The granular endoplasmic reticulum is, for the most part, arranged in layers. The mitochondria appear short with a dense matrix and distinct transverse crests. There is a very large Golgi and a considerable amount of smooth-surfaced endoplasmic reticulum. A number of slightly irregular, moderately stained secretory granules occupy an extensive juxtanuclear area and are particularly numerous in the very large Golgi zone. One of the larger granules has a dense nucleoid (arrow). A number of much smaller, dense granules can be found throughout the cytoplasm. The corticotrope, or ACTH-producing cell, has been reported to contain granules which change in appearance under varying stimuli. In the normal rat the small dense granules are infrequent. It has been suggested that the corticotropes may be cells of the follicular type which have been transformed into cells with an increased number of secretory granules. This corticotrope was found adjacent to a group of cells in a follicular arrangement and may be one of such a group. 26,500×

2. The thyrotrope is a distinctive cell of angular or polygonal shape. Dark elongated mitochondria stand out against a pale background cytoplasm. These cells have a very scanty, granular endoplasmic reticulum. The cytoplasm contains a number of vesicles and microtubules. The granules of the thyrotrope are the smallest to be identified in any of the cells of the pars distalis; most are less than 0.15 μ in diameter. Most of the granules are found near the border, but, among the few in the interior of the cytoplasm, one finds occasional larger and elongated granules. A dense layer is seen on the external surface of the plasma membrane in an invagination of the cell border (arrow). The granules of the gonadotrope (Gt) are larger than those of the thyrotrope. 18,000×

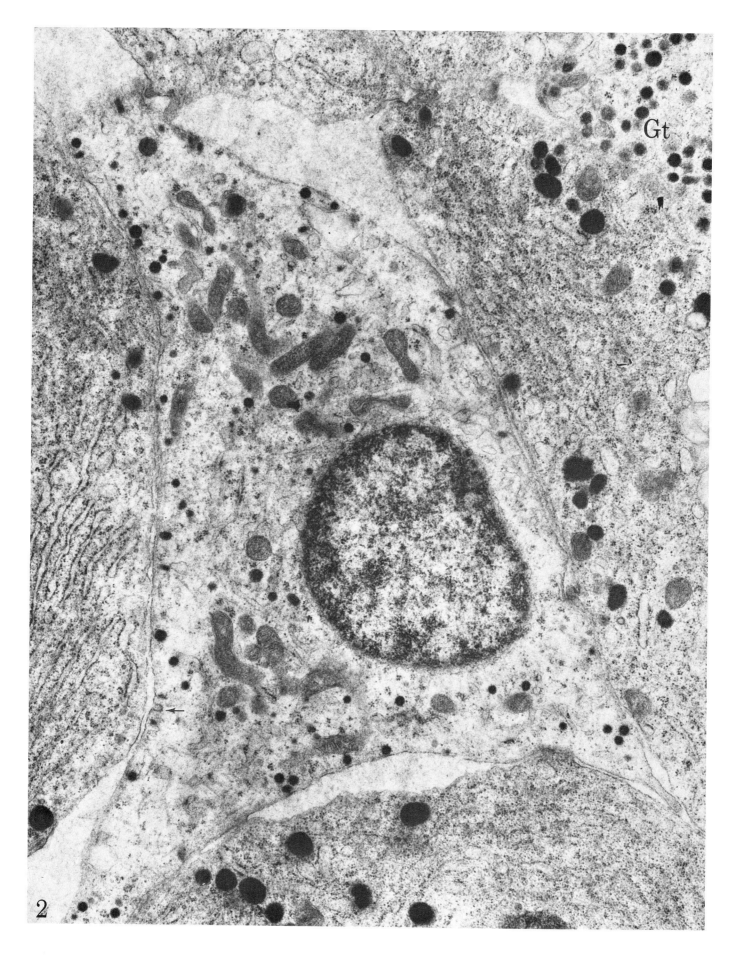

Gt

2

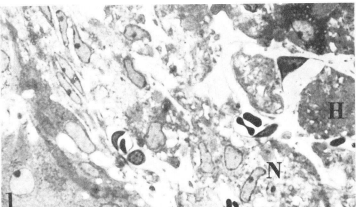

Pars Nervosa—The Hypophysis

1. L. In the pars nervosa (N) large, extremely irregular pituicytes with pale cytoplasm and elongated nuclei are embedded in an extensive network of unmyelinated nerve fibers. Among the nerve fibers are more densely stained masses of cytoplasm termed Herring bodies (H). These are accumulations of large nerve endings filled with secretory material. The small group of myelinated nerve fibers at a considerable distance from the pars intermedia is an uncommon finding. The pars nervosa is highly vascularized. 600×

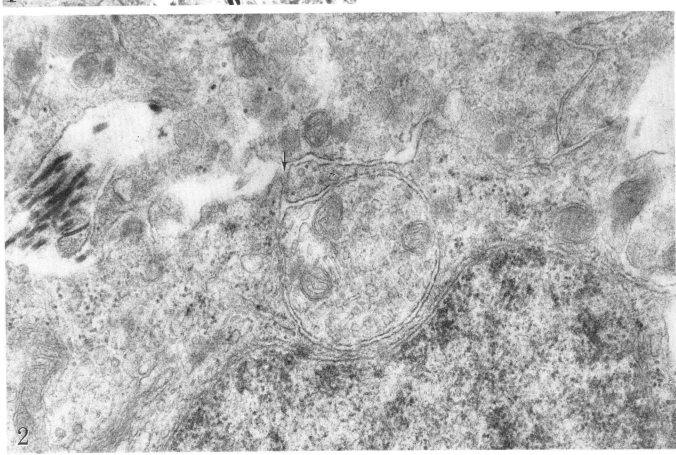

2. A nerve ending is embedded in a pituicyte. The infolding of the plasma membrane (arrow) of the pituicyte is typical of that seen in the Schwann cell–axon relationship. The nerve ending contains a large number of elongated or irregular synaptic vesicles and a few mitochondria. Other nerve endings with both small, synaptic-type vesicles and larger, moderately dense granules are in contact with the pituicyte, but they are not embedded in its cytoplasm. 45,000×

3. A Herring body is composed of a number of bulbous nerve endings, most of which are filled with large neurosecretory granules and a few smaller vesicles. Some endings contain groups of smaller, dense glycogen granules. A number of cytoplasmic microtubules are seen in the nerve endings as well as in the unmyelinated nerve fibers among them. Some of these cytoplasmic microtubules have a dense luminal content. Long cytoplasmic processes of pituicytes extend between the nerve endings in the Herring body. A clearly defined circumferential lamina surrounds the Herring body but does not surround the individual endings. A loosely arranged layer of collagen is interposed between the Herring body and the basal lamina of the capillary. The nerve endings in the pars nervosa are the terminals of neurosecretory cells in the hypothalamus. 50,000×

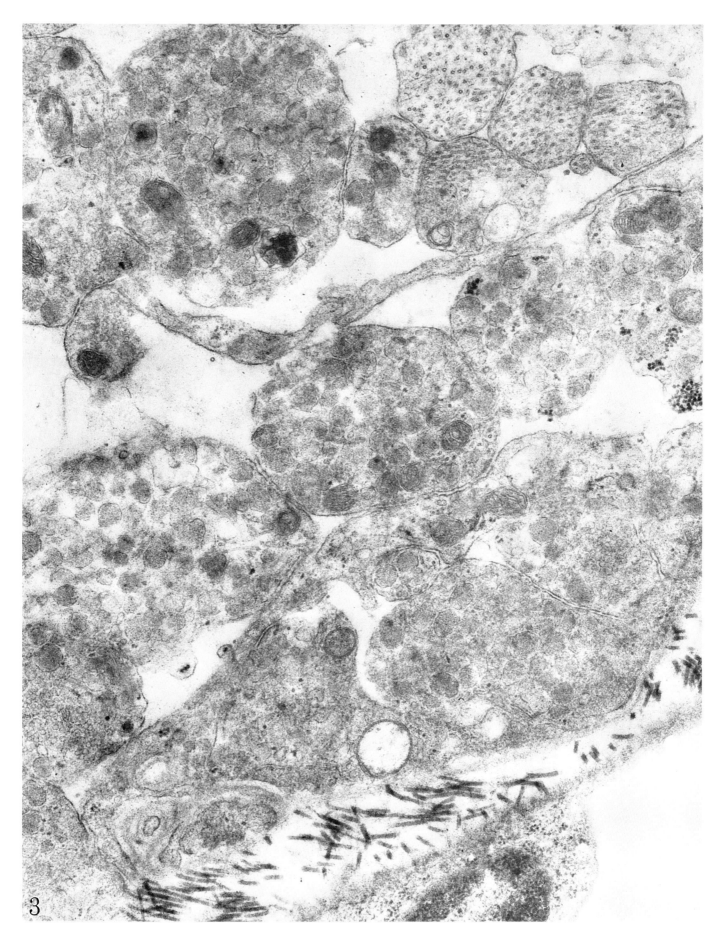

3

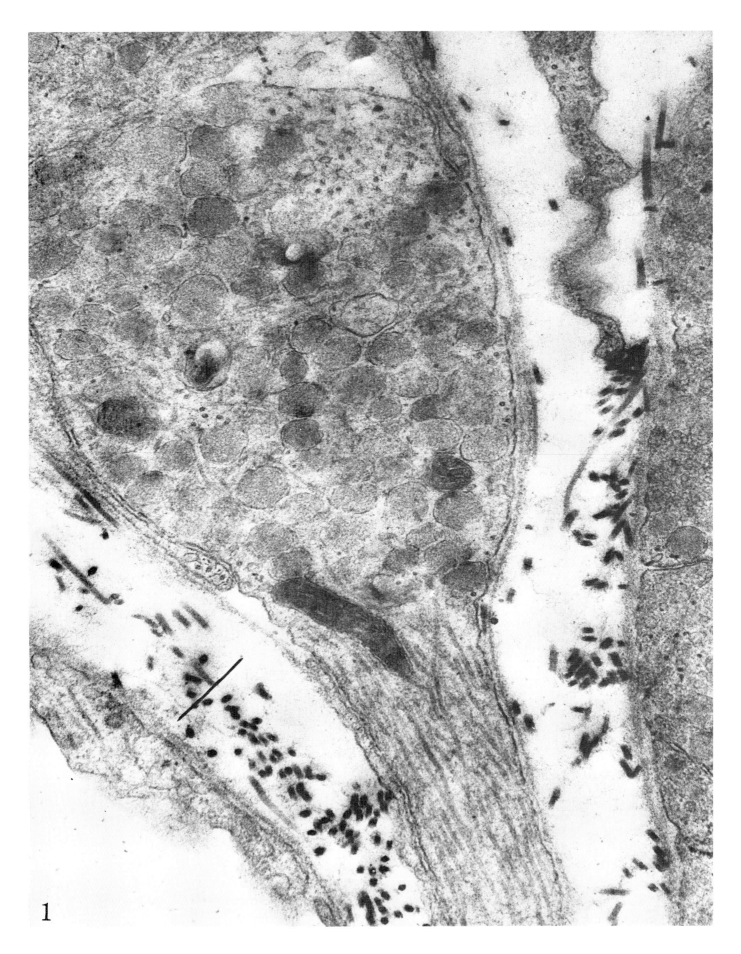

1

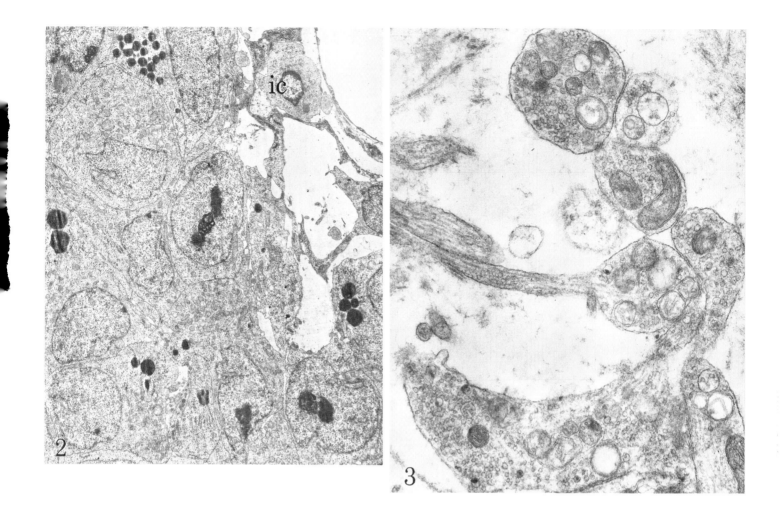

The Pars Nervosa—The Hypophysis

1. This axon of a neurosecretory cell of the hypothalamus terminates on a fenestrated capillary in the pars nervosa. Thin pituicyte cytoplasmic processes are applied to the nerve ending as widely separated plaques. The neurosecretory ending and the applied pituicyte processes are included in a circumferential lamina. The unmyelinated axon has a very high concentration of cytoplasmic microtubules. A high percentage of these cytoplasmic microtubules are heavily stained. The elongated mitochondrion with longitudinally oriented cristae has an extremely dense matrix. Within the bulbous ending the neurosecretory granules vary considerably in their density and each is surrounded by a well-defined membrane. Many cytoplasmic microtubules are found between the granules and in a localized border area. Transversely sectioned microtubules in this area show a marked variation in the stain of the content of the central lumen. 72,000×

The Pineal Body

The functions of the pineal gland have only recently begun to be recognized. Experimental evidence indicates that pineal hormones under some conditions have an inhibitory effect on gonadal development, probably through an indirect influence on some sites in the hypothalamus or the hypophysis.

2. This organ is made up of cords of parenchymal epithelioid cells, the pinealocyte or chief cells, vascular channels with large perivascular spaces, and occasional interstitial cells. The pinealocyte nuclei are large and pale with large, dense, often multiple, nucleoli. Both large and small dense granules are sharply defined in the pinealocyte. The interstitial cells (ic) have small, dense nuclei in contrast with those of the pinealocyte. Pinealocyte nuclei are elongated and often deeply indented. One nucleus contains a prominent principal nucleolus in which the nucleolonema is more distinct and two very dense accessory nucleoli in which a lesser amount of nucleolonema is discernible. This nucleolar arrangement is similar to that seen in the Sertoli cell of the testis. Long cytoplasmic processes extend into the capaceous perivascular space. The density of the lipid droplets in this photograph has been preserved by a prolonged dehydration in acidified ethanol. 2350×

3. The thin cytoplasmic prolongations of the pinealocytes end in bulbous or semilunar-shaped endings. The process has a substructure typical of an unmyelinated axonal nerve process with many cytoplasmic microtubules, tubular endoplasmic reticulum, and mitochondria occupying the majority of the cytoplasm. The bulbous or semilunar endings are filled with clear vesicles 500–600 Å in diameter and larger vesicles with moderately stained, homogeneous content. The microtubules extend into these endings from the cytoplasmic processes. (Dehydrated for 24 hours in ethanol at approximately pH 7.3.) 30,000 ×

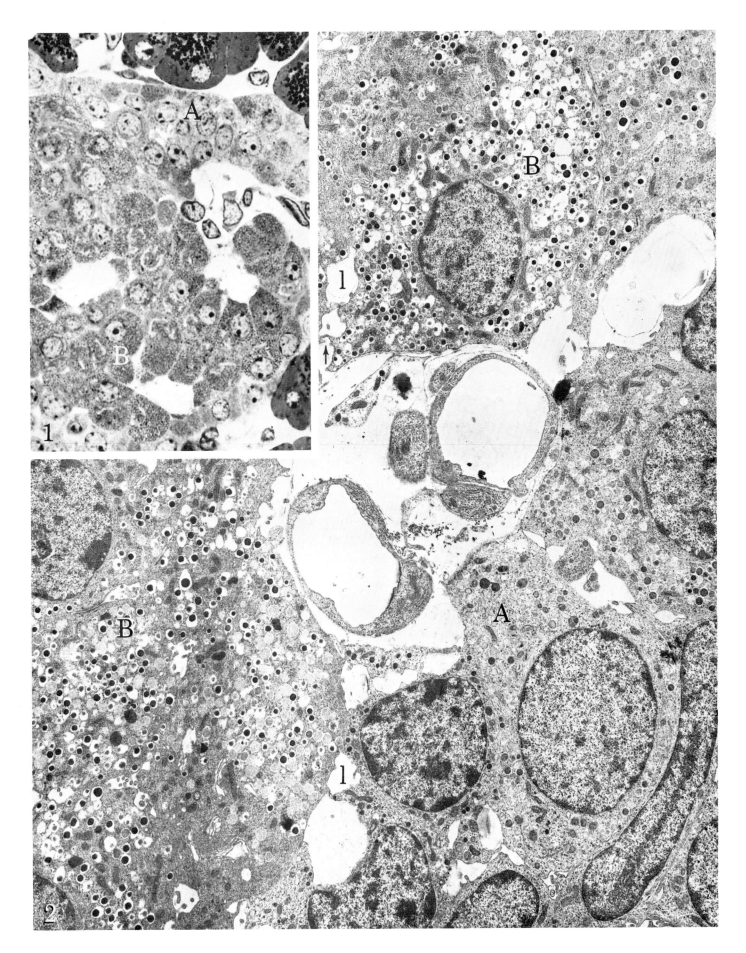

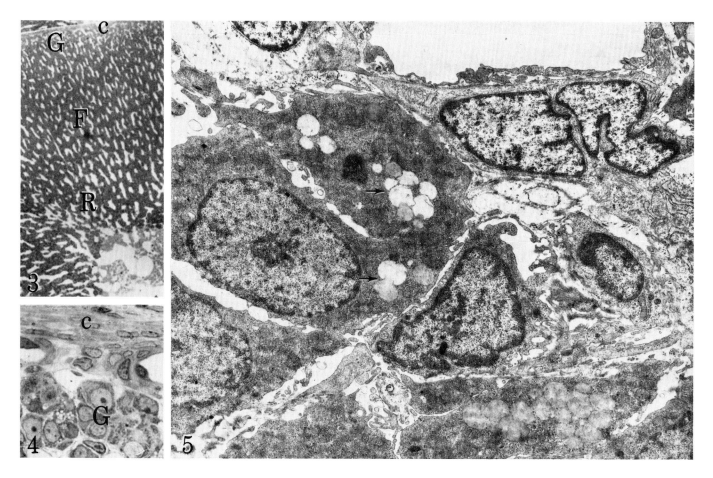

The Pancreatic Islets

The pancreatic islets (of Langerhans) are located among acini of the exocrine pancreas. These glands of internal secretion have a controlling influence on the storage and mobilization of carbohydrates in the body. One hormone, insulin, is released into the bloodstream in response to increases in blood sugar level. This hormone dictates the storage of carbohydrates as glycogen or fat in cells throughout the body. Another product of the islet is glucagon, which stimulates the reconversion of glycogen to glucose in order to satisfy the energy requirements of the body. Different types of cells have been distinguished in the islets. The main types are the beta cells, which synthesize insulin, and the alpha cells, which produce glucagon.

1. L. By the toludine blue method utilized for this work, the beta (B) cells appear darker and their granules more distinct. They occupy a large part of the islet. The alpha (A) cells are seen at the periphery. There is an abundant capillary supply. 680×

2. A pericapillary space intervenes between capillaries and the surrounding beta (B), and alpha (A) cells. The beta cells, which are most numerous, contain dense granules surrounded by a halo and, in addition, a number of larger dense bodies. The cytoplasm varies in density, and a prominent Golgi apparatus is evident in one cell. In the alpha cells the granules are homogeneous and vary considerably in density, but they are considerably less dense than those of the beta cells. A wide range of density of granule can be found in any one alpha cell. Microvilli project into intercellular lacunae (l), some of which open (arrow) into the pericapillary space. Cilia have been reported to project into these lacunae in several species of animal. 7000×

The Suprarenal (Adrenal) Glands

The suprarenal gland is an endocrine gland composed of two main parts, a cortex and a medulla. The cortex is of coelomic mesothelial origin, while the medulla is derived from the neural crest of the ectoderm along with the other autonomic ganglia.

3. L. Beneath a thin fibrous tissue capsule (c) the cortex is made up of cords of cells which stain quite intensely with toluidine blue. It is subdivided into a glomerulosa (G), a fasciculata (F), and a reticularis (R). These zones are identified by their arrangement in relation to the blood spaces between the cords of cells as well as by some differences in the cells. The medulla (M) is made up of large, pale cells. 160×

4. L. At a higher magnification the fibrous capsule (c) contains numerous fibroblasts and a large number of vascular channels immediately adjacent to the glomerulosa (G). The glomerulosa is made up of moderately large cells with pale nuclei which contain prominent nucleoli and abundant, dense, but vesiculated cytoplasm. 600×

5. The capillaries in the capsule, adjacent to the glomerulosa, are of the fenestrated type. Fibroblasts, with irregular nuclei, and a loose arrangement of collagen fibrils are located between the capsular vessels and the glomerulosa. The cells in this superficial part of the glomerulosa have densely stained mitochondria, many of which have transversely oriented crests, vacuoles, and granular inclusions of varying density. The moderately stained, lipid-like content of some granules protrudes into the clear vacuole in several instances (arrows). Many cytoplasmic processes protrude from the surfaces of the glomerular cells. 6000×

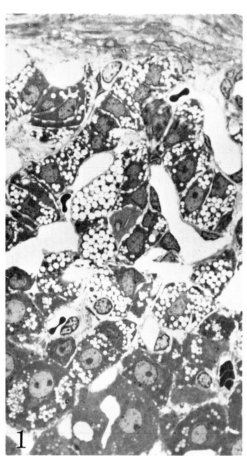
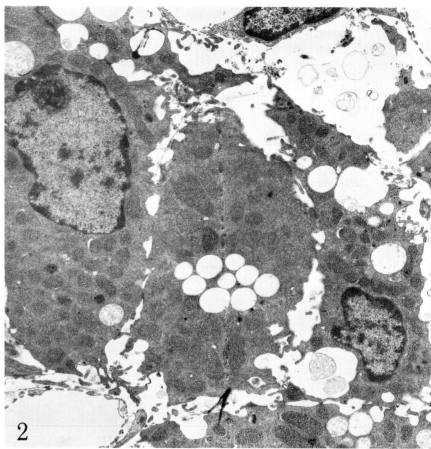

The Suprarenal (Adrenal) Glands

1. L. Capsule, glomerulosa, and a small part of the fasciculata of the suprarenal gland. The cords of cells are bordered by endothelium-lined blood sinusoids, some of which contain a few red blood cells. The cells of the glomerulosa appear vacuolated. This vacuolation decreases at the border of the fasciculata. A few dense granules can be seen in some of the glomerulosa cells. 680×

2. The cells of the glomerulosa have a dense cytoplasm, an irregular nucleus with dense peripheral chromatin, and a prominent nucleolus. The mitochondria have a very dense background, and all of the crests appear tubular in the deeper levels of the glomerulosa. Small dense granules and larger vacuoles, some with a flocculent material, are seen in the cytoplasm. In tissues fixed with osmium tetroxide alone most of the vacuoles in the glomerulosa contain lipid. These lipids are lost in Epon-embedded, aldehyde- and osmium-fixed cells unless special procedures are employed to preserve them during dehydration. 6250×

The Suprarenal Gland

3. L. The cords of the fasciculata give way to a network which, in certain planes of section, appear as sheets of cells in the reticularis. The cells of the fasciculata have circular, pale nuclei.

These cells are frequently seen as single cords between the sinusoids. Lipid droplets, some of which have a pale central region, are quite numerous in the fasciculata and decrease in number in the reticularis. The density of the lipid droplets in this photograph has been preserved by a prolonged dehydration in acidified ethanol. 800×

4. The single row of cells in the fasciculata is bordered by sinusoids with a fenestrated endothelial cell lining. A few long, curved cytoplasmic projections from the fasciculata cells project into the perisinusoidal spaces. Lighter, irregular cytoplasmic processes from an interstitial cell (ic) are seen in some places in the perisinusoidal space. Some prolongations of the cytoplasm extend between the cells of the cord of the fasciculata. The cells of the fasciculata have rounded nuclei, a large number of homogeneous lipid droplets, and a smaller number of dense granules. The cells have a dense cytoplasm with numerous mitochondria, all of which have the tubular type of cristae. 2700×

5. In a section of the reticularis the cells differ only slightly from the fasciculata except in the increased number of dense granules. In many of these cells, very large mitochondria (lm) are seen. Some of the mitochondria contain a number of large dense granules. Between the mitochondria there is a large amount of endoplasmic reticulum. As in the fasciculata, some lighter interstitial cell cytoplasm (ic) can be seen between the endothelium and the cells of the reticularis. 5000×

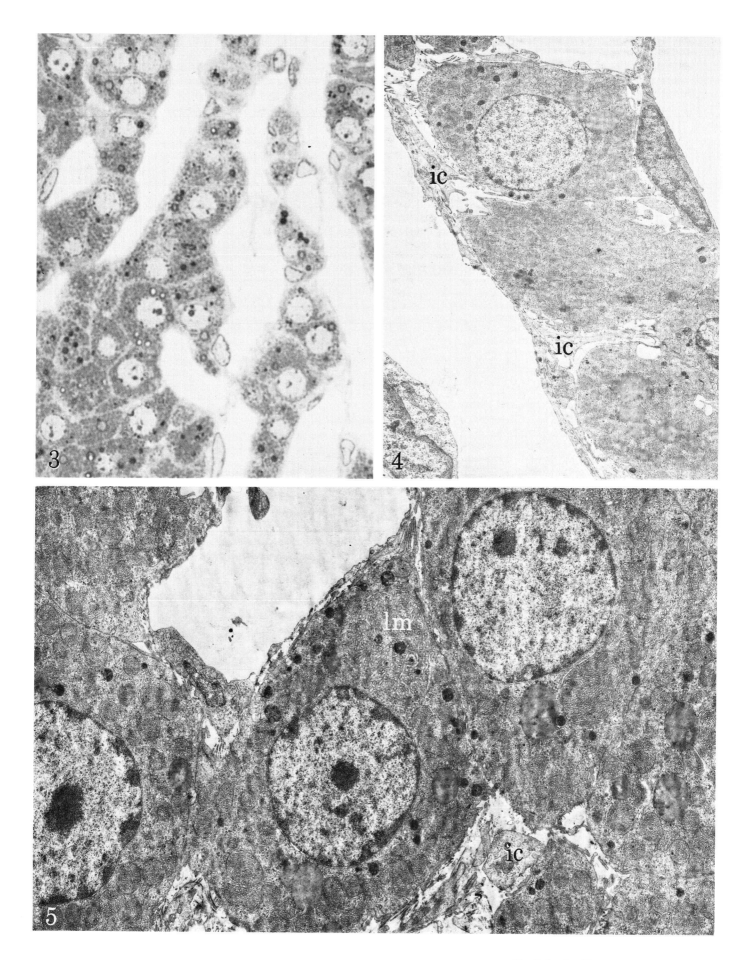

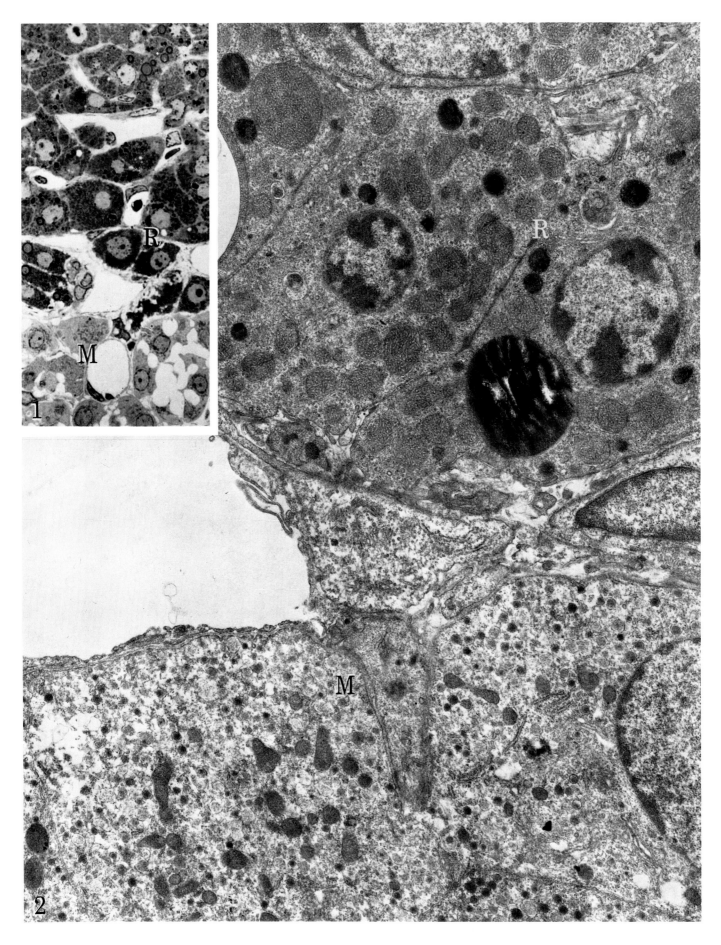

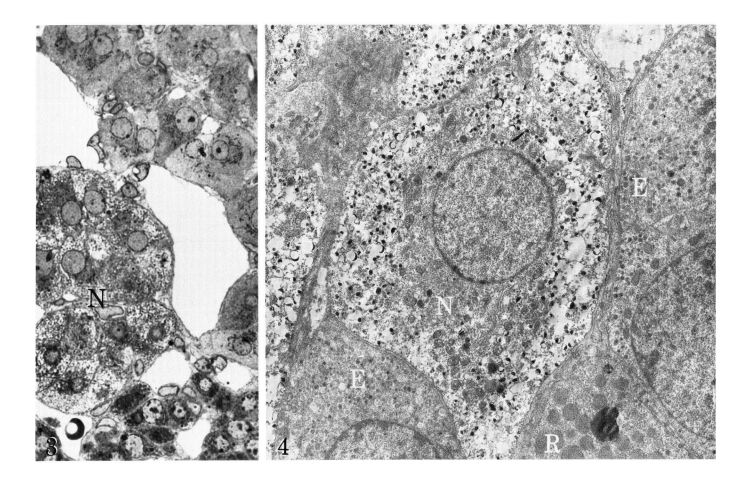

The Suprarenal Gland—The Corticomedullary Junction

1. L. The cytoplasm of cells in the reticularis (R) of the adrenal cortex contains a number of lipid droplets and dense granules. The cells of the medulla (M) have a much paler background cytoplasm with numerous, very small, faintly defined granules in some of the cells. 600×

2. Dehydration in acidified alcohol results in an extreme density of lipid droplets in the reticularis (R). A number of small dense granules are found throughout the cytoplasm. The medullary cells (M), derived from neuroectoderm, contain a large number of moderately dense granules. This type of granule is believed to be the epinephrine or adrenaline storing granule of these neurosectory cells. These are chromaffin cells of the adrenal medulla. The cells have densely stained, slender mitochondria, a quantity of granular endoplasmic reticulum, and an extensive juxtanuclear Golgi apparatus. Interstitial cells and fibrils occupy the extracapillary space in the adrenal medulla as well as in the cortex. 8500×

The Suprarenal Medulla

3. L. In light microscopy of thin sections a distinct difference can be seen in two types of cells of the suprarenal medulla. The cytoplasm of one type (N) appears vacuolated with fine, dense superimposed granules. These are believed to be the cells which synthesize and store norepinephrine. The other cell type is more homogeneous with smaller, less dense granules and a darker cytoplasmic background. These are believed to be the sites of synthesis and storage of epinephrine. The mitochondria of both types of cells stain intensely. 680×

4. In the type of suprarenal medullary cell (N) which is believed to store norepinephrine, there is a vesiculated cytoplasm with extremely dense content filling a part of most of the vesicles. The perinuclear cytoplasm is much more dense than the peripheral. In contrast, the epinephrine storing cells (E) have a large number of moderately dense cytoplasmic granules. A cell from the reticularis (R) of the adrenal cortex contains a large, dense lipid droplet. 5000×

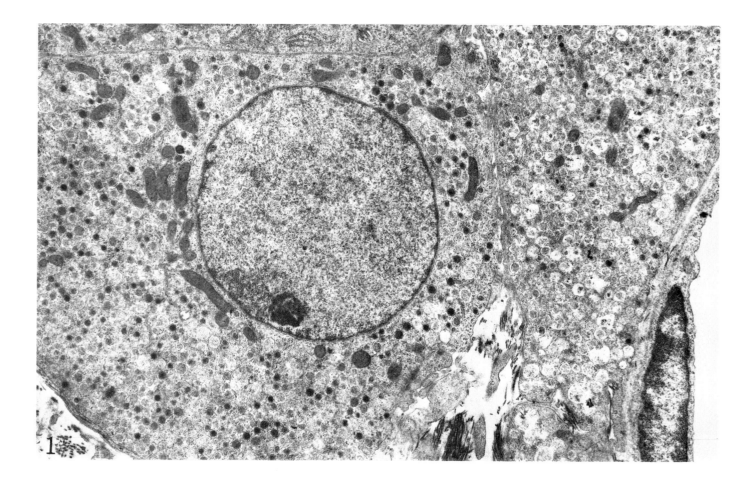

The Suprarenal Medulla

1. The mitochondria are slender and densely stained in the cells wherein the moderately dense granules are believed to store epinephrine. The secretory granules vary greatly in their intensity of stain. The ratio of dense to less dense granules varies from cell to cell. 7500×

2. Nerve endings (n) on the cells which are believed to store epinephrine appear to take the form of demilunes, which form a cap over one pole of the cells. These endings contain a large number of moderately stained small vesicles and some denser, larger granules. In these cells, the Golgi apparatus is located in the juxtanuclear region toward the nerve ending. Large dense granules, in addition to the numerous smaller granules, are seen in the cytoplasm of these cells. A higher concentration of the smaller, moderately dense granules is found toward the capillary pole (c). 7000×

The stimulation of these cells in emergency situations by the nerve endings results in the release of epinephrine. This, in turn, stimulates the sympathetic nervous sytem. The result is an increased heart rate and cardiac output and an increased basal metabolic rate. It affects a mobilization of carbohydrates stored in the liver. The epinephrine also indirectly affects the adrenal cortex through its stimulation of the hypophysis to release ACTH.

3. An enlargement of a part of Fig. 2 demonstrates the long area of contact between the preganglionic fiber terminal and the epinephrine-producing cell. Typical synaptic plaques occur only in localized areas. The small vesicles seem more numerous around these synaptic plaques, while the large vesicles with dense content are found toward the periphery of the flattened terminal. A number of these larger vesicles have a dense content, while others are empty or contain a small peripheral or central accumulation of dense material. In the cytoplasm of the receptor, irregularly shaped vesicles contain a moderate amount of density, while smaller vesicles have densely stained content. The material in the cisternae of the granular endoplasmic reticulum is moderately stained. 21,000×

4. A very large, rounded nerve ending with numerous synaptic vesicles is in synapse with the norepinephrine storing cells. Most of the vesicles are small, from 400 to 600 Å in diameter, with very little staining of the content. Only a few larger, more dense granules can be seen. An inclusion (arrow) has the appearance of a moderately stained lipid droplet. Dense synaptic plaques can be seen along the junctional zone. 23,000×

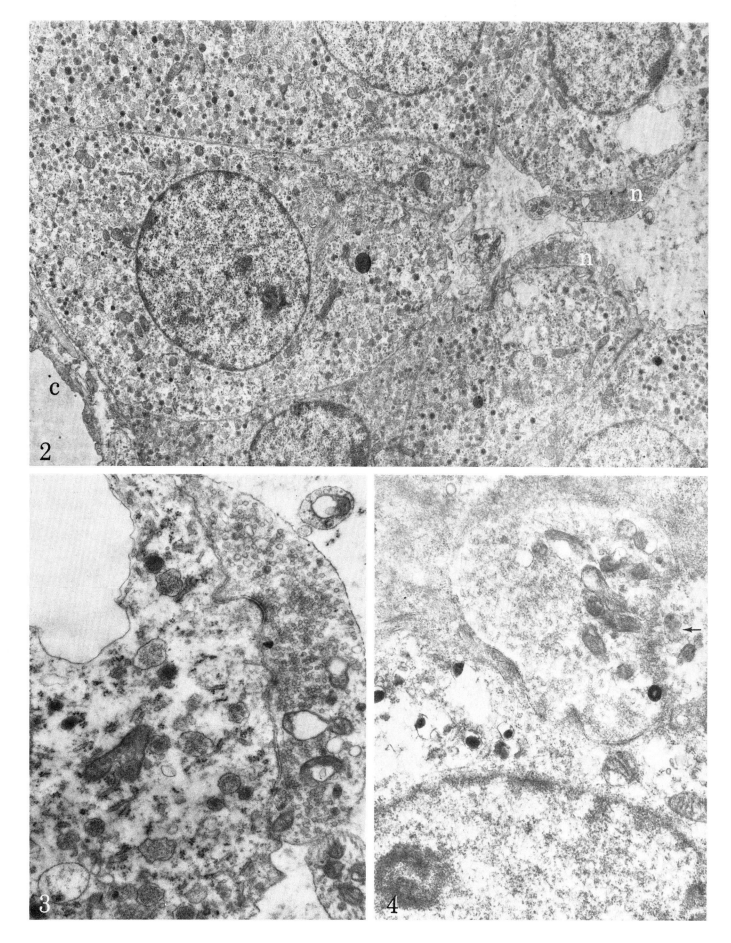

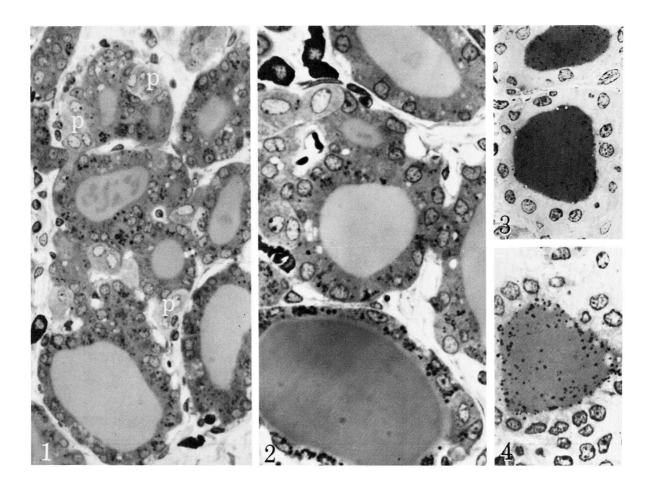

The Thyroid Gland

The thyroid gland is an unusual type of endocrine gland wherein the maturation process of the synthesized product takes place within closed follicular lumina outside of the cells. These follicular cells synthesize and/or transport the precursors of the thyroglobulin to the follicular lumen. After maturation of the synthesized product in the follicular lumen, the same cells must transport and secrete this hormone in a utilizable form into the extracapillary space.

1. L. Follicular epithelial cells which vary considerably in height surround the colloid-filled lumina. These cells, in this adult animal, contain a number of dense lysosomal granules. Large, pale parafollicular (p) cells are seen at the base of the follicular cells in many areas. The parafollicular cells are more numerous in the apical poles of the lobes of the thyroid gland. 440×

2. L. At a higher magnification, in addition to the dense granules, pale vacuoles can be seen in many of the thyroid follicular cells. These pale globules represent droplets of colloid which have been shown to be in transport toward the base of the cell. At this magnification a few dense granules, smaller than the lysomal granules of the follicular cells, can be seen in the parafollicular cells. 700×

3. L. The glycoprotein thyroglobulin is iodinated in the follicular lumen. Light microscopic autoradiography can be utilized to demonstrate the rate of transfer and the localization of isotopes within an organ. Two minutes after the intravenous injection of ^{125}I, its incorporation into the thyroglobulin is indicated by the developed silver grains which are found almost exclusively over the colloid. 800×

4. L. At 10 minutes after the injection of ^{125}I, there is an increase in radioactivity over the colloid, but there is still little evidence of tracer over the cell. 800×

5. The thyroid follicular cell is a cuboidal cell with a moderate number of microvilli which project into the colloid filled lumen. The nucleus contains coarse, densely stained chromatin. The cisternae of the granular endoplasmic reticulum are dilated The cell has multiple large Golgi complexes in the supranuclear region. Large dense granules, which increase in frequency with the age of the animal, are found in the cytoplasm of the follicular cells. These granules, which are present from the base to the apex of the cells, have been shown to contain acid phosphatase and other enzymes common to lysosomes. The mitochondria are elongated and many are aligned along the lateral borders of the cell. Capillaries with fenestrated endothelium are closely applied to the follicular epithelial cells. Large, dense granules are seen in the cytoplasm of the endothelial cells. Basal laminae accompany the follicular epithelial and the capillary endothelial cells. 17,500×

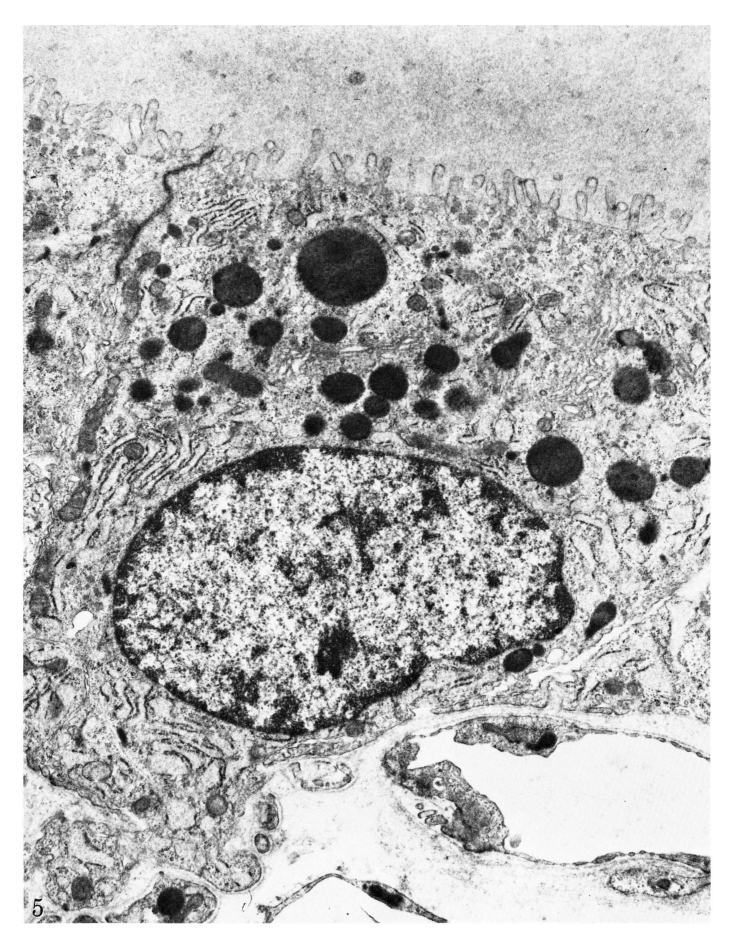

5

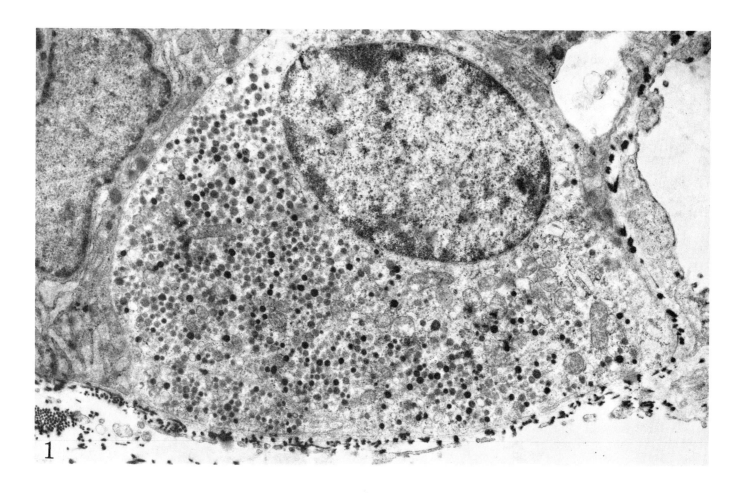

1

The Thyroid Gland

1. The parafollicular cell, described many years ago, is closely applied to the thyroid follicular cell but does not reach the lumen. It is placed between the basal lamina and the follicular epithelial cells. The parafollicular cell has a much less dense background cytoplasm than the follicular cells. A very large number of small granules of variable density are found in the cytoplasm. These granules measure from 1500 to 2500 Å in diameter. It has been reported recently that a high cholinesterase activity is present at the periphery of these cells. There is considerable evidence to indicate that they are the source of the hypocalcemic hormone calcitonin (thyrocalcitonin) and they have therefore been termed C-cells. It has been suggested that a monoamine such as a catecholamine may be present in the granules. These cells appear to be derived from the ultimobranchial body. 11,000×

The Parathyroid Glands

2. L. In the rat, the parathyroid glands are embedded within the thyroid gland. These parathyroids appear as clusters of large pale cells with irregular nuclei and a few small granules in their cytoplasm. The densely packed cells are arranged as cords around groups of capillaries. In this young animal only one type of cell, the principal or chief, was found. 600×

3. In the parathyroid of the perfused animal, a moderate amount of intercellular space is very commonly seen. The cells are large and have irregular nuclei. The cytoplasm is quite dense and contains a number of vacuoles and dense granules. The cords of cells are bordered by a basal lamina, and the neighboring capillaries are of the type with a fenestrated endothelium. In the pericapillary space, which is considerable, the collagen fibrils are widely scattered. 3250×

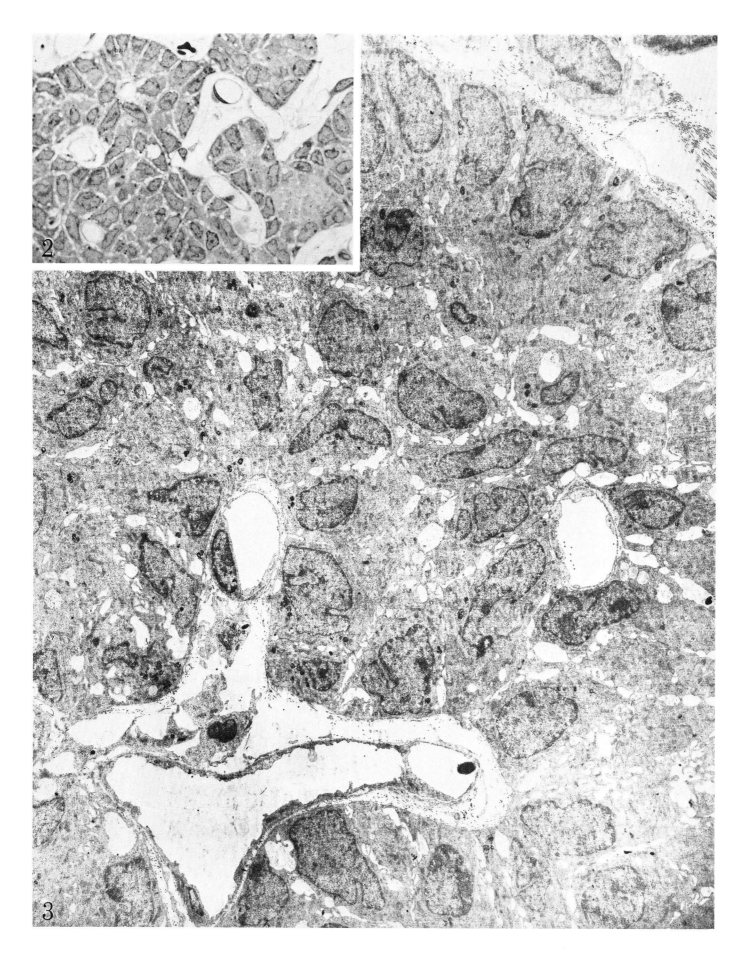

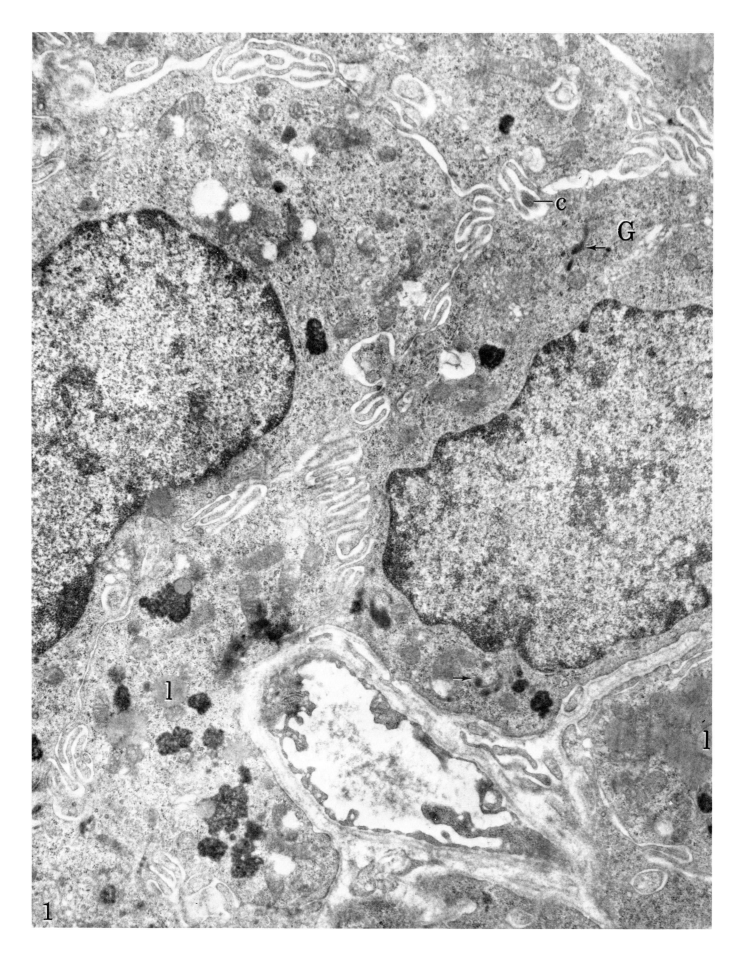

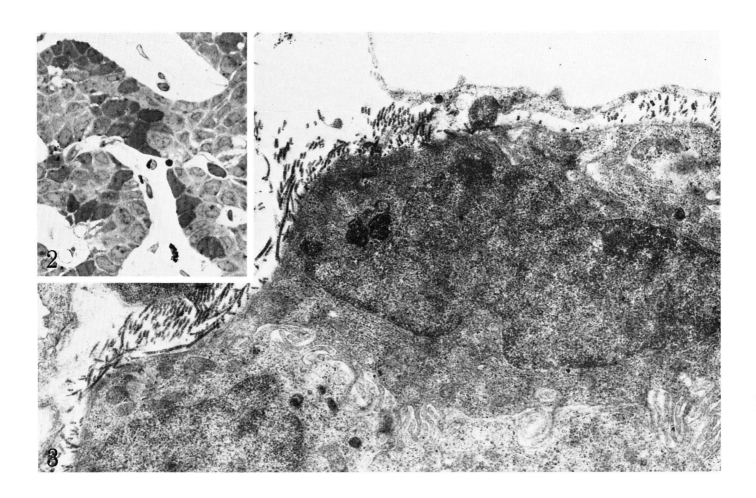

The Parathyroid Glands

1. The borders of the parathyroid chief cell are quite folded, and the folds interdigitate with similar processes of neighboring cells. Cytoplasmic processes project out into the pericapillary space. The cells contain a large number of ribosomes, varying numbers of lipid droplets (l), and a number of lipofuscin granules. Dense content is seen in flattened sacs of the endoplasmic reticulum and in expansions at the ends of these membranous profiles (arrows). A moderate-sized Golgi complex (G) surrounded by a number of vesicles is located in the pole more distant from the capillary. Moderately dense granules and elongated dense tubules or sacs are found in this zone. A cilium (c) is seen in the intercellular space. 36,000×

2. L. The parathyroid gland of an adult rat exhibits a number of cells which stain considerably more densely than do the principal cells. These darker, oxyphil cells are often found in small groups but are also seen singly among the lighter principal cells. 560×

3. An oxyphil cell with relatively scanty cytoplasm is located between lighter principal cells. The cytoplasm of the former contains a higher concentration of granular endoplasmic reticulum than does the principal cell. Both types of cell contain dense granules. 14,000×

REFERENCES

NEUROSECRETION

Bargmann, N. (1966). Neurosecretion. *Intern. Rev. Cytol.* **19**, 183–201.

Lever, J. D. (1964). Structural aspects of endocrine secretion. *Arch. Biol. (Liege)* **75**, 436–452.

HYPOPHYSIS

Barnes, B. G. (1962). Electronmicroscope studies on the secretory cytology of the mouse anterior pituitary. *Endocrinology* **71**, 618–628.

Barnes, B. G. (1963). The fine structure of the mouse adenohypophysis in various physiological states. *In* "Cytologie de l'adénohypophyse" (J. Benoit and C. Da Lage, eds.), pp. 91–103. C.N.R.S., Paris.

Benoit, J., and C. Da Lage, eds. (1963). "Cytologie de l'adénohypophyse," Colloq. Intern. Centre Natl. Rech. Sci. No. 128. C.N.R.S., Paris.

Duffy, P. E., and Menefee, M. (1965). Electron microscopic observations of neurosecretory granules, nerve and glial fibers and blood vessels in the median eminence of the rabbit. *Am. J. Anat.* **117**, 251–286.

Farquhar, M. G., and Rinehart, J. F. (1954). Electron microscopic studies of the anterior pituitary gland of castrate rats. *Endocrinology* **54**, 516–541.

Farquhar, M. G., and Rinehart, J. F. (1954). Cytologic alterations in the anterior pituitary gland following thyroidectomy: An electron microscope study. *Endocrinology* **55**, 857–875.

Farquhar, M. G., and Rinehart, J. F. (1955). Further evidence for the existence of two types of gonadotrophs in the anterior pituitary of the rat. *Anat. Rec.* **121**, 394.

Green, J. D. (1966). Electron microscopy of the anterior pituitary. *In* "The Pituitary Gland" (G. W. Harris and B. T. Donovan, eds.), Vol. I, pp. 233–241. Butterworth, London and Washington, D. C.

Herlant, M. (1964). The cells of the adenohypophysis and their functional significance. *Intern. Rev. Cytol.* **17**, 299–382.

Kurosumi, K. (1968). Functional classification of cell types of the anterior pituitary gland accomplished by electron microscopy. *Arch. Histol. Japon.* **29**, 329–362.

Monroe, B. G. (1967). A comparative study of the ultrastructure of the median eminence, infundibular stem and neural lobe of the hypophysis of the rat. *Z. Zellforsch. Mikroskop. Anat.* **76**, 405–432.

Monroe, B. G., and Scott, D. E. (1966). Ultrastructural changes in the neural lobe of the hypophysis of the rat during lactation and suckling. *J. Ultrastruct. Res.* **14**, 497–517.

Purves, H. D. (1961). Morphology of the hypophysis related to its function. *In* "Sex and Internal Secretions" (W. C. Young and G. W. Corner, eds.), 3rd ed., Vol. 1, pp. 161–239. Williams & Wilkins, Baltimore, Maryland.

Siperstein, E. R. (1963). Identification of the adrenocorticotrophin-producing cells in the rat hypophysis by autoradiography. *J. Cell Biol.* **17**, 521–546.

Smith, R. E., and Farquhar, M. G. (1966). Lysosome function in the regulation of the secretory process in cells of the anterior pituitary gland. *J. Cell. Biol.* **31**, 319–347.

PINEAL GLAND

Duncan, D. (1966). Notes on the fine structure of the pineal organ of cats. *Texas Rept. Biol. Med.* **24**, 576–587.

Wartenberg. H., and Gusek, W. (1965). Licht-und Elektronenmikroskopische Beobachtungen über die Struktur der Epiphysis Cerebri des Kaninchens. Prog. Brain. Res. (J. Ariëns-Kappers and J. P. Schadé, eds.), Vol. 10; pp. 296–316. Elsevier, Amsterdam.

Wolfe, D. E. (1965). The epiphyseal cell: An electron microscopic study of intercellular relationships and intracellular morphology in the pineal body of the albino rat. Prog. Brain Res. (J. Ariëns-Kappers, and J. P. Schadé, eds.), Vol. 10, pp. 332–386. Elsevier, Amsterdam.

Wurtman, R. J., Axelrod, J., and Kelly, D. E. (1968). "The Pineal." Academic Press, Inc., New York.

PANCREATIC ISLETS

Bencosme, S. A. (1955). The histogenesis and cytology of the pancreatic islets in the rabbit. *Am. J. Anat.* **96**, 103–152.

Björkman, N., Hellerström, C., Hellman, B., and Petersson, B. (1966). The cell types in the endocrine pancreas of the human fetus. *Z. Zellforsch. Mikroskop. Anat.* **72**, 425–445.

Caramia, F. (1963). Electron microscopic description of a third cell type in the islets of the rat pancreas. *Am. J. Anat.* **112**, 53–64.

Kern, H. F. (1970). The fine structure of pancreatic α-cells under normal and experimental conditions. Reprinted from The Structure and Metabolism of the Pancreatic Islets, Wenner-Gren Symposium No. 16. Edited by S. Falkmer, B. Hellman and I. B. Täljedal, Pergamon Press, Oxford and New York.

Kobayashi, K. (1966). E. M. studies of the Langerhans islets in the toad pancreas. *Arch. Histol. Japon.* **26**, 439–482.

Like, A. A. (1967). The ultrastructure of the islets of Langerhans in man. *Lab. Invest.* **16**, 937–951.

Meyer, J., and Benscome, S. A. (1965). The fine structure of normal rabbit pancreatic islet cells. *Rev. Can. Biol.* **24**, 179–205.

Munger, B. L., Caramia, F., and Lacy, P. E. (1965). The ultrastructural basis for identification of cell types in the pancreatic islets. II. Rabbit, dog and opossum. *Z. Zellforsch. Mikroskop. Anat.* **67**, 776–798.

SUPRARENAL GLAND

Brenner, R. M. (1966). Fine structure of adrenocortical cells in adult male rhesus monkeys. *Am. J. Anat.* **119**, 429–454.

Elfvin, L. G. (1967). The development of the secretory granules in the rat adrenal medulla. *J. Ultrastruct. Res.* **17**, 45–62.

Wood, J. G. (1963). Identification of and observations on epinephrine and norepinephrine containing cells in the adrenal medulla. *Am. J. Anat.* **112**, 285–304.

Zelander, T. (1959). Ultrastructure of mouse adrenal cortex. *J. Ultrastruct. Res.* Suppl. 2, 1–111.

THYROID

Ekholm, R. (1967). Thyroglobulin biosynthesis in the rat thyroid, *J. Ultrastruct. Res.* **20**, 103–110.

Ekholm, R., and Ericson, L. E. (1968). The ultrastructure of the parafollicular cells of the thyroid gland in the rat. *J. Ultrastruct. Res.* **23**, 378–402.

Ekholm, R., and Smeds, S. (1966). On dense bodies and droplets in the follicular cells of the guinea pig thyroid. *J. Ultrastruct. Res.* **16**, 71–82.

Van Heyningen, H. E., and Sandborn, E. B. (1963). Sites of iodine-binding in rat thyroid as shown by radioautography with I^{125}. *J. Appl. Phys.* **34**, 2525.

PARATHYROID

Capen, C. C., and Rowland, G. N. (1968). The ultrastructure of the parathyroid glands of young cats. *Anat. Record* **162**, 327–340.

Davis, R., and Enders, A. C. (1961). Light and electron microscope studies on the parathyroid gland. *In* "The Parathyroids" (R. O. Greep, and R. V. Talmage, eds.), pp. 76–93. Thomas, Springfield, Illinois.

Lever, J. D. (1965). Fine structural organization of the human and rat parathyroid glands. *In* "The Parathyroid Glands" (P. J. Gaillard, R. V. Talmage, and A. M. Budy, eds.), pp. 11–17. Univ. of Chicago Press, Chicago, Illinois.

Mazzocchi, G., Meneghelli, V., and Frasson, F. (1967). The human parathyroid glands: An optical and electron microscopic study. *Sperimentale* **117**, 383–447.

Trier, J. S. (1958). The fine structure of the parathyroid gland. *J. Biophys. Biochem. Cytol.* **4**, 13–22.

Wetzel, B. K., Spicer, S. S., and Wollman, S. H. (1965). Changes in fine structure and acid phosphatase localization in rat thyroid cells following thyrotropin administration. *J. Cell Biol.* **25**, 593.

SPECIAL SENSES
The Eye and the Ear

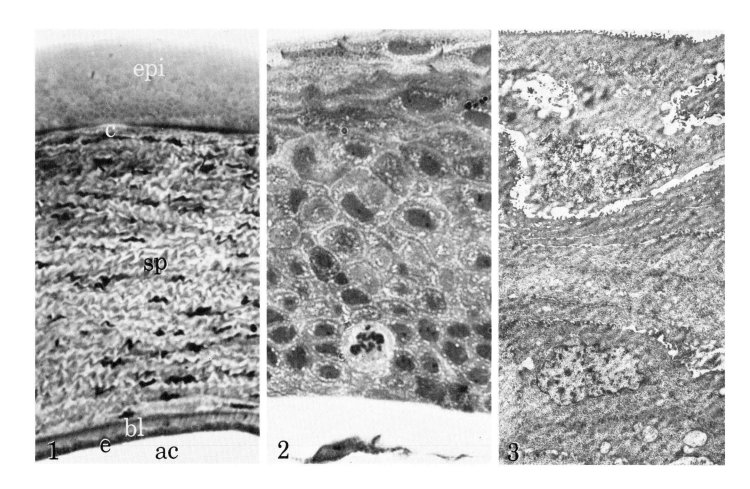

Cornea

1. L. The cornea, which is the transparent portion of the outer fibrous tunic of the eyeball, is made up of five layers. The outer layer is stratified squamous epithelium (epi). Beneath this is a thin limiting layer of fine collagen (c). This has been named, from its light microscopic appearance, Bowman's membrane. The thick substantia propria (sp) is made up of collagen and fibroblasts in a large number of undulating layers. Flattened fibroblasts stain densely in comparison with the intervening collagen. A more or less homogeneous layer, a well-developed basal lamina (bl) (Descemet's membrane), separates the substantia propria from the endothelial lining (e) of the fluid-filled anterior chamber (ac) of the eye. 180 ×

2. L. At a higher magnification the corneal epithelium can be seen to be made up of a number of layers of cells which are outlined by intercellular spaces. A cell in mitosis is seen in the germinal layer of the epithelium. 800 ×

3. Superficial epithelial cell layers are extremely flattened. There is evidence of degeneration of the nuclei of the more superficial cells. The intercellular spaces widen toward the external surface of the cornea. 3500 ×

4. The fibrils of collagen, parallel to one another within a layer of the substantia propria, are oriented at right angles to those

in alternate layers. A fibroblast is located between layers of collagen fibrils. 3500 ×

5. When the cornea is sectioned obliquely to the collagen fibrils, a mosaic pattern is obtained. 3500 ×

6. At a higher magnification of these obliquely sectioned layers, the superimposed fibrils from neighboring layers give a netlike appearance. 25,000 ×

7. L. The inner limiting (il) (or Descemet's) layer is a fairly broad band intervening between the fibrous substantia propria (sp) and the endothelial (e) lining of the anterior chamber of the eye. 600 ×

8. The internal limiting layer, a basal lamina between the substantia propria and the endothelial cells, is composed of a homogeneous matrix interlaced by collagen fibrils and more densely stained elastic fibers. The concentration of the fibers and the fibrils is greater in the outer portion of the thick lamina. The thick endothelium of the cornea appears as a layer of low cuboidal cells. The intercellular borders are quite folded. The nuclei of these cells vary considerably in form, and one contains a large, dense nucleolus. A moderate quantity of granular endoplasmic reticulum and a few dense granules are seen in the cytoplasm. The mitochondria (arrows) are quite indistinct. A pair of centrioles is located in the portion of the cytoplasm bordering on the anterior chamber (ac) of the eye. 7800 ×

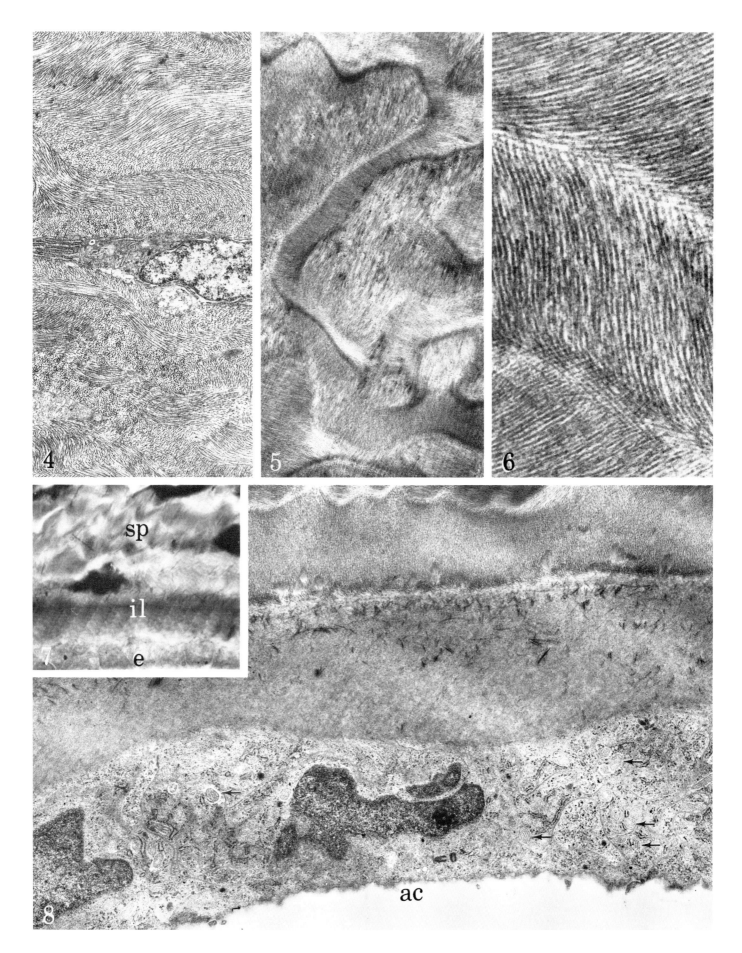

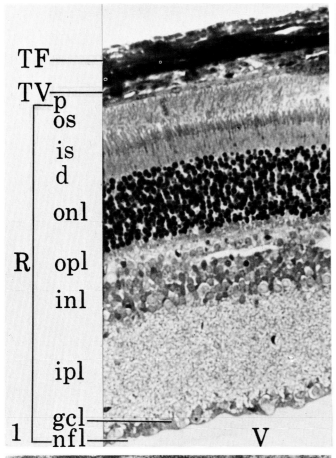

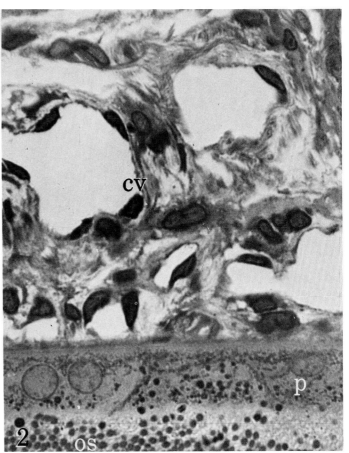

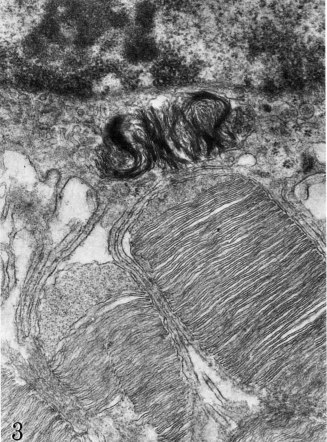

The Eye

1. L. The coats of the eyeball in the light-sensitive area are subdivided into: (1) an outer layer, the tunica fibrosa or (sclera) (TF), which is a dense fibrous tissue coat; (2) a middle coat, the tunica vasculosa (TV), or the choriovascular layer; and (3) the retina (R). The retina is the light-sensitive portion and is made up of a pigmented cell layer (p), which is not neuronal in nature, and nervous tissue which is organized into well-ordered layers. These nervous layers include, from the exterior to the interior of the retina: the rod and cone outer segment (os); the rod and cone inner segments (is); the desmosome layer (d), commonly termed the outer limiting membrane; the outer nuclear layer (onl) of the rods and cones; the outer plexiform layer (opl); the inner nuclear layer (inl) of bipolar cells and other relay neurons and supporting cells; the inner plexiform layer (ipl); the ganglion cell layer (gcl); the nerve fiber layer (nfl), axons of the ganglion cells which will form the optic nerve, and an internal limiting lamina (membrane). The light must pass from the vitreous body (V) through all of the layers of the retina to the outer segments of the rods and cones. From these light-sensitive, outer segments the stimulus is transmitted back to the rod and cone cell bodies, is relayed to the bipolar cells, then to the ganglion cells, and is transmitted to the optic nerve. 200 ×

2. L. A plexus of blood vessels in loose connective tissue with large interstitial cells of the choriovascular (cv) coat is separated from the single layer of pigmented cells (p) by a band of uniform width. The pigmented cells have large, rounded, pale nuclei and a number of dense granules in the cytoplasm. Some of these granules are found among the numerous, less dense, transversely sectioned rod outer segments (os). 800 ×

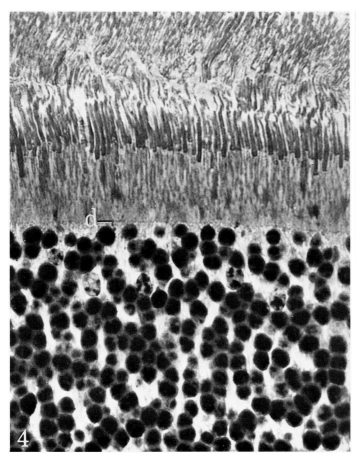

Retina

3. The tips of the retinal rod outer segments are capped by a crescent of granular material. Proximal to this material the membranes are arranged in horizontal layers which, in one rod, traverse the whole width of the segment. In the adjacent rod they appear interrupted. Both the granular tip and the membrane folds are included within the plasma membrane of the rod cell. A layer of cytoplasm from the pigment cell clothes each rod outer segments. Thus two cytoplasmic processes are seen between the receptors. A dense laminated inclusion (a myeloid body) is seen in the pigment cell. 30,000 ×

4. L. Rod cells with extremely dense nuclei far outnumber the cone cells. Several of the latter are clearly identified by the discrete densities in larger, pale nuclei. The layer of inner segments of the rods and cones is separated from the nuclear layer by a thin, almost continuous line. This appearance in light microscopy led to the terminology "outer limiting membrane," but it is, in actuality, a band of desmosomes (d) which occurs between supportive and receptor cells. 720 ×

5. The line of demarcation between the inner segment and the nuclear layers in the electron micrograph is seen as an interrupted band of large desmosomes. These desmosomes are between the pale cytoplasm of glial or supporting "Müller cells" and the dense cytoplasm of the inner segments at their junctions with rod and cone cell bodies. The numerous slender cytoplasmic bodies of individual rod cells and fewer, thicker processes of the cone cells have been sectioned longitudinally, obliquely, and transversely between the nuclei in this outer nuclear layer. The large, pale nuclei of cone cells with distinct chromatin masses are in sharp contrast with the much denser and smaller nuclei of the rods. 4500 ×

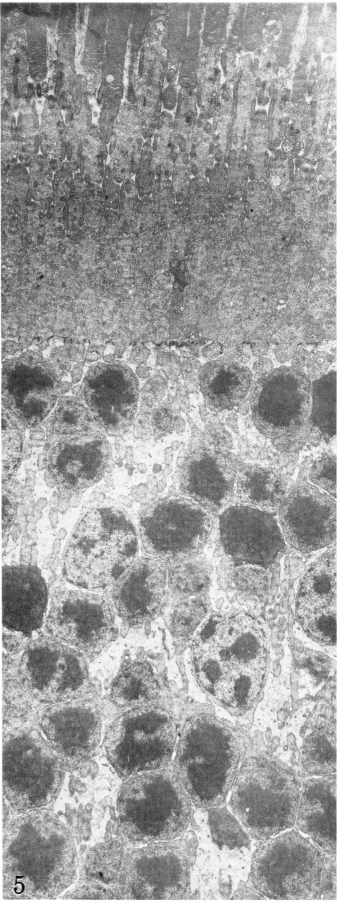

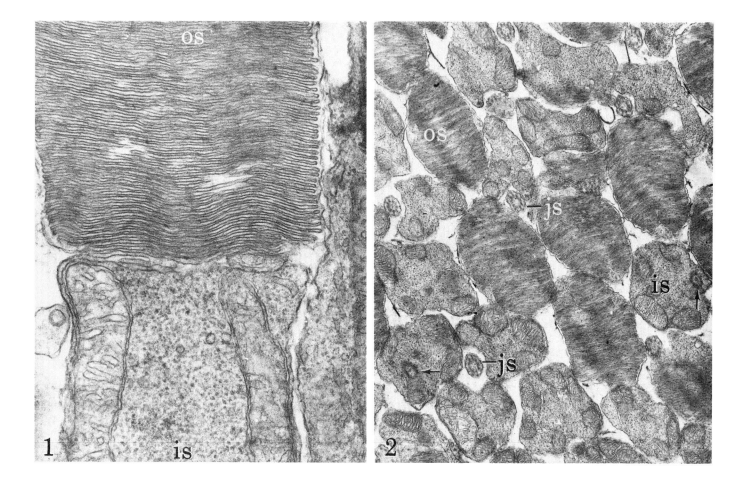

Retina

1. At the junction between the outer (os) and the inner segments (is) of the rods a distinct change of character occurs. In this view the outer segment, with its parallel membrane folds, appears completely separated from the inner segment wherein mitochondria are located at the periphery. 50,000 ×

2. In a transverse section through this junctional zone the outer segments (os), the inner segments (is) with their peripheral mitochondria, and the junctional, cilium-like stalks (js) are demonstrable. Centrioles or basal bodies are sectioned within the inner segments (arrows). It is difficult to obtain perfect transverse sections of both inner and outer segments in one field since the segments are often joined at an angle. 15,000 ×

3. In part of an inner segment (is) of a rod and in the junctional zone of another the transversely striated root of the cilium-like process and the extension into the outer segment (os) are seen in detail. The rootlet is made up of tightly packed, longitudinally oriented subunits with narrow, alternate lighter and darker transverse striations. In the less dense bands these subunits can be clearly distinguished, while in the darker bands two almost continuous lines cross the rootlet (white arrows). A number of cytoplasmic microtubules parallel the rootlet in the inner segment. In the cilium-like stalk the doublets appear to be associated with filaments which branch out in several directions (arrows). The microtubules give way to filaments which appear to reach the narrower stack of membrane profiles within the outer segment. Other filaments radiate out to the wider of the two stacks of membranes. Typical, lateral filamentous bridges in this case appear to reach the plasma membrane of the stalk. 58,500 ×

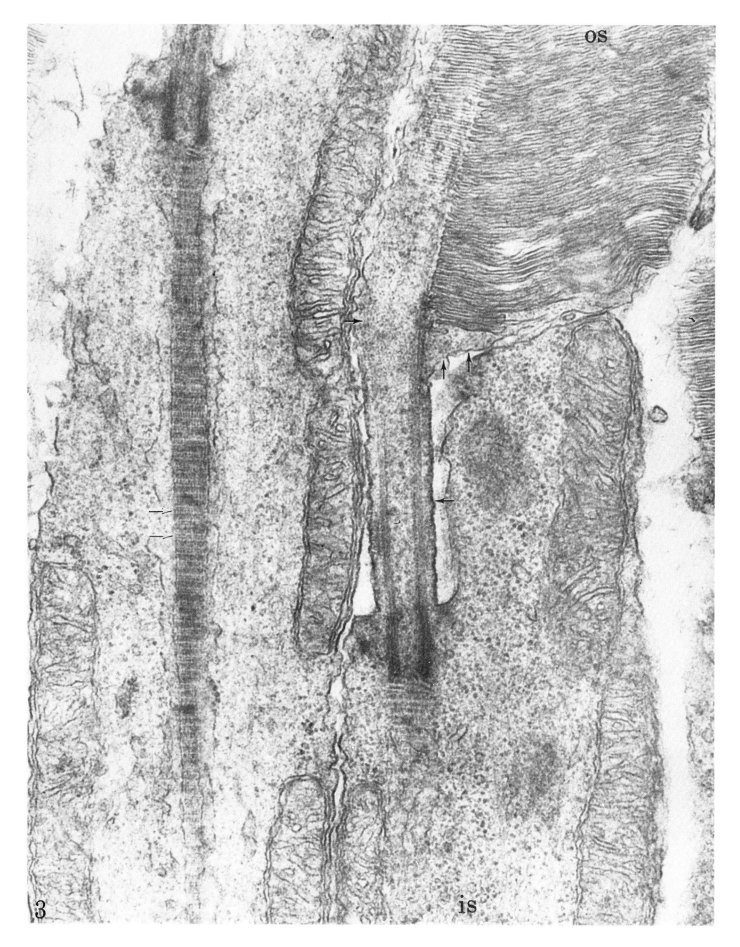

os

is

3

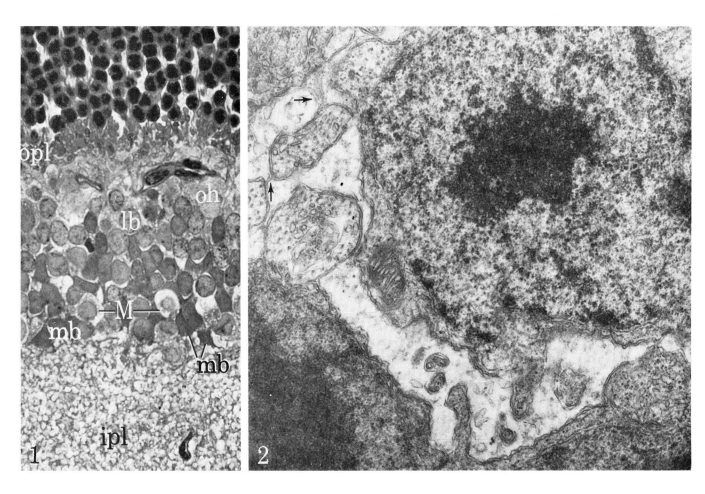

Retina

1. L. In the retina of the rat the outer plexiform layer (opl) between the denser outer nuclear and the lighter, inner nuclear layers is quite narrow. This layer has an abundant supply of capillaries among numerous nerve fibers and large synaptic nerve endings. Five cell types are identified in the inner nuclear layer: (1) large, pale cells, the outer horizontal association cells (oh), which are probably involved in relays between rod and cone cells in different parts of the retina; (2) the midget or monosynaptic bipolar cell (mb), the small neurons with dense nuclei and cytoplasm which synapse in the outer plexiform layer with single cone cell processes and with ganglion cell processes in the inner plexiform layer (ipl); (3) the larger or polysynaptic bipolar cells (lb), which form synapses with both rod spherules and cone foot plates in the outer plexiform layer and with ganglion cell processes in the inner plexiform layer; (4) inner horizontal association cells, similar in appearance to the outer; and (5) cells with clear cytoplasm, the Müller (M) glial cells, from which cytoplasmic processes extend through, and envelop, the cells of the outer nuclear layer. These processes form desmosomal junctional complexes with receptor cells at the point where the fiber is in continuity with its inner segment. Other Müller cell cytoplasmic processes extend through the inner plexiform layer. 600 ×

2. A cone cell nucleus and its narrow band of cytoplasm are surrounded by the lighter Müller cell cytoplasm. The latter contains widely scattered organelles, including vesicles or tubes of endoplasmic reticulum and transversely and obliquely sectioned microtubules and filaments. The rod fibers are in-cluded in invaginations of the plasma membrane of the Müller cell in exactly the same fashion as the Schwann cell envelops unmyelinated axons (arrows). Every rod fiber contains a considerable number of transversely sectioned cytoplasmic microtubules and a smaller number of larger tubules of agranular reticulum. 27,000 ×

3. Rod spherules (rs), many of which contain one mitochondrion, are packed with synaptic vesicles. The large, irregular pyramidal terminal expansion of a cone (ct) is packed with similar vesicles against a darker background. Both types are in contact with and encompass, within deep invaginations or furrows of their surfaces, small dendritic processes from cells of the inner nuclear layer. Very pale Müller cell cytoplasm surrounds the rod spherules and the cone expansions and extends into deep invaginations of their surfaces, where some appear as clear vacuoles. Some interreceptor contacts can be seen between rod spherules and the cone expansions. An interreceptor branch of a cone expansion which contains synaptic vesicles can be seen, surrounded by Müller cell cytoplasm within the invagination of one spherule. Dense osmiophilic ribbons bordered by a zone of closely packed synaptic vesicles are found in both types of receptor terminals. 20,000 ×

4. An enlargement of the osmiophilic ribbon in a rod spherule from Fig. 1 shows three trilaminar membrane-like profiles in parallel. The central one is irregular and appears interrupted. Filamentous cross bridges appear to interconnect these parallel structures. 150,000 ×

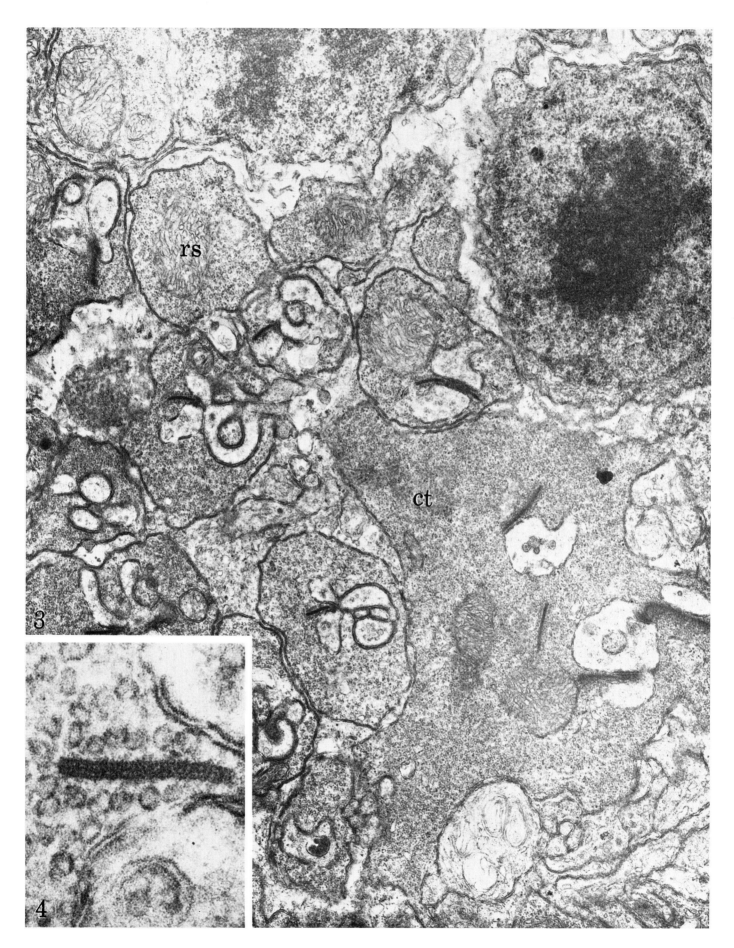

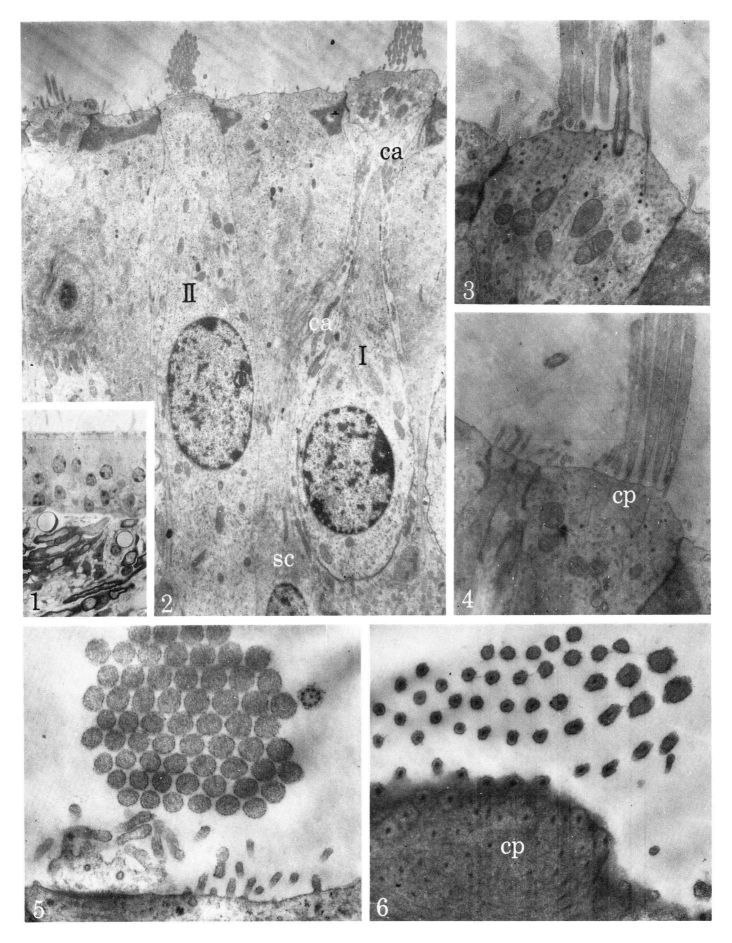

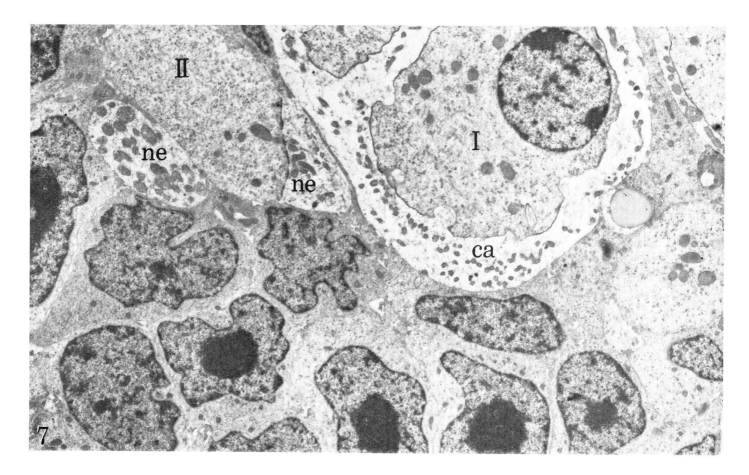

The Inner Ear—The Crista Ampullaris

Sensory epithelium is found in the organ of Corti and in the maculae of the utricule and saccule and in the cristae ampullares of the vestibular labyrinth. Two basic cells types are found in the organ of Corti and in the crista. These are the hair cells and their supporting cells. Stimuli received by the hairs on the sensory cells are transmitted by synapses to terminal expansions of the afferent nerve fibers.

1. L. Crista ampullaris. The hairs are seen as tufts projecting from tall columnar cells. Supporting cells, with their small dark nuclei at a deeper level than those of the sensory cells, contain dense masses abutting against the apexes of the sensory cells. This is part of a markedly developed apical cell web.

2. The most extensive expansion of the afferent nerve endings is the calyx (ca) in association with the type I (I) hair cell of the crista ampullaris. Here the calyx reaches to the level of the narrow neck of the sensory cell. The resulting appearange is that a "bottle shaped" cell, the majority of which is engulfed by the afferent nerve expansion. The type II (II) sensory hair cells synapse with smaller expanded nerve endings at its base. The hairs are actually one kinocilium associated with a group of specialized stereocilia on each sensory cell. The apexes of supporting cells contain dense masses of packed fibrillar material adjacent to intercellular junctional complexes. The smaller, darker nuclei of the supporting cells (sc) are at a deeper level than those of the hair cells.

3. A section has intersected the single kinocilium which arises from a basal body in the apex of each hair cell. A longitudinally oriented central density in the basal part of a stereocilium extends as a rootlet for some distance into the cell body. Dense granules can be seen along the rootlet. Numerous vertically oriented cytoplasmic microtubules are seen in the cytoplasm.

4. A longitudinal view of the stereocilia demonstrates a narrow basal attachment and an expanded distal portion. Each of the stereocilia contains a basal central density. Extensions of these densities appear as rootlets which pierce a slightly fibrillar area, the cuticular plate (cp) in the apex of the sensory cell.

5. In transverse section, the distal parts of stereocilia have a moderately stained background. The single kinocilium of the cell is also seen in transverse section.

6. A transverse section through the proximal parts of a group of stereocilia demonstrates the appearance of a central dense circular body as the section approaches the cell proper. In the cell apex the rootlets become enmeshed in the cuticular plate (cp).

7. The calyx (ca), a greatly expanded afferent nerve terminal, in transverse section encircles the type I sensory cell body (I). The type II (II) cell is in contact with two separate expansions (ne) of another afferent fiber. The nuclei of the supporting cells are quite irregular in outline.

REFERENCES

THE EYEBALL

Bairati, A., Jr., and Orzalesi, N. (1963). The ultrastructure of the pigment epithelium and of the photoreceptor-pigment epithelium junction in the human retina. *J. Ultrastruct. Res.* 9, 484–496.

Bernstein, M. H. (1961). Functional architecture of the retinal epithelium. *In* "The Structure of the Eye" (G. K. Smelser, ed.), pp. 139–150. Academic Press, New York.

de Robertis, E., and Lasansky, A. (1961). Ultrastructure and chemical organization of photoreceptors. *In* "The Structure of the Eye" (G. K. Smelser, ed.), pp. 29–49. Academic Press, New York.

Jakus, M. A. (1961). The fine structure of the human cornea. *In* "The Structure of the Eye" (G. K. Smelser, ed.), pp. 343–366, Academic Press, New York.

Miller, W. H. (1960). Visual photoreceptor structures. *In* "The Cell" (J. Brachet, and A. E. Mirsky, eds.), Vol. 4, pp. 325–364. Academic Press, New York.

Sjöstrand, F. S. (1965). The synaptology of the retina. *Ciba Found. Symp. Colour Vision, Physiol. Exptl. Psychol.* pp. 100–151.

THE INNER EAR

Engström, H. The innervation of the vestibular sensory cell. Acta Otolarynogologica, Supp. *163*:30, 1960.

Engström, H. Wersall, J. The ultrastructural organization of the organ of Corti and of the vestibular sensory epithelium. Exper. Cell Res. Supp. *5*:460, 1958.

Iurato, S., 1967. Submicroscopic Structure of the Inner Ear. Pergamon Press, Oxford, 367 pp.

Kimura, R. S., 1966. Hairs of the cochlear sensory cells and their attachment to the tectorial membrane. Acta Otolaryng., 61:55–72.

Kimura, R. S., H. F. Schuknecht, and I. Sando, 1964. Fine morphology of the sensory cells in the organ of Corti of man. Acta Otolaryng., 58:390–408.

Lim, D. Three dimensional observation of the inner ear with the scanning electron microscope. Acta Otol. Supp. *225*, 1969.

Wersäll, J. Studies of the structure and innervation of the sensory epithelium of the cristae ampullaris in guinea pigs. Acta Otolaryngologica, Supp. *126*.

SUBJECT INDEX

A

Absorption
  of colloidal gold, 103
  of ferritin, 109
Acrosome, 146, 149
ACTH-producing cell, 170
Alveolus, 29, 30
Ameloblast, 102
Aorta, 96
Apparatus, juxtaglomerular, 132, 133
Arteriole, 96
Astrocyte, 67
Autoradiography, 184
Axon, 5, 65
  myelinated, 5, 65
  unmyelinated, 5, 69

B

Bladder, urinary, 140, 141
Body
  chromatoid, 146, 147
  pineal, 175
Bone, 43–46
Bone marrow, 46–51

C

C-cell, thyroid gland, 186
Canaliculus
  intercellular, 105
  intracellular, 106
Cap, acrosomal, 146–149
Capillary, 29, 98
Capsule, Bowman's, 135
Cartilage, 40–42
Cell web, apical, 109, 201
Cells, 1–23
  absorptive, 106–111
  acinar, submaxillary gland, 119
  alpha, 177
  argentaffin, 104, 113
  argyrophil, 104, 114
  basal, 127
  beta, 177
  centroacinar, 121
  chief
    parathyroid gland, 186–189
    pineal, 175
    stomach, 104–107
  cholinesterase rich argyrophil, 114
  delta, 177
  endothelial, 96–99
  enterochromaffin, 104
  epinephrine storing, 181
  follicular, thyroid, 184
  germinal, 145
  glial, 67, 70
  goblet, 107, 110, 111
  intercalated duct, 119
  interstitial of Leydig, 145
  littoral, 126
  mesangial, 130
  in mitosis, 17–19, 107, 115, 123
  mucous neck, 104
  myoepithelial, 165
  neurosecretory, 172, 175
  norepinephrine storing, suprarenal gland, 181, 182
  oxyntic, 104
  oxyphil, parathyroid gland, 189
  pancreatic acinar, 7–9, 119–121
  Paneth, 113
  parafollicular, thyroid gland, 184, 186
  parietal, 104–106
  principal, parathyroid gland, 186–189
  red, blood, see Erythrocyte
  Sertoli, 145, 146
  stellate (of Kupffer), 123, 126
  sustentacular, 145
  theca luteal, 162
Centriolar satellite, 16
Centriole, 16, 19
Chromatin, 5
Chromosomes, 17, 115
Cilia, 20, 30, 164
Circulatory system, 95–100
Collagen, 33–35, 45
Connective tissue, 33–39
Cornea, 192
Corpus luteum, 160, 162
Corpuscle, renal, 130, 132
Corticotrope, 170
Crista ampullaris, 202, 203
Crypt
  duodenal, 106, 113
  ileum, 113
Cytochemistry, 11
Cytochrome c reductase, localization of reaction product, 11
Cytoplasmic microtubule, see Microtubule, cytoplasmic

D

Dendrite, 69
Desmosome, 15, 54
  septate, 15
Droplet, colloid, thyroid gland, 184
Duct
  biliary, 127
  granulated, salivary gland, 119
  intercalated, salivary gland, 119
  pancreatic, 121
  striated, salivary gland, 119
Ductus deferens, 155
Ductus epididymus, 153
Duodenum, 106–109, 115

E

Ear, inner, 202, 203
Elastic fibers, 33, 37
Elastic lamina, 96
Enamel, 102, 103
Endometrium, 163

Mossy fiber, 69
Multivesicular body, 7
Muscle, 75–88
  cardiac, 86, 87
    atrial, 87
    conducting fiber, 87
    granules in, 86, 87
    intercalated disc, 86, 87
    ventricular, 86
  skeletal striated, 81–85
    sarcoplasmic reticulum, 84
    T tubule, 84
    triad, 81, 84
  smooth, 76–79, 141
    innervation, 79
Myelin sheath, 5, 6, 65
Myelocyte
  basophilic, 51
  eosinophilic, 51
  heterophilic, 51
Myofilament, 77, 81, 82

N

Nephron, 130
Nerve ending, 70, 72, 201
Nerve peripheral, 5, 6, 65
Nervous system, 61–74
Neuromuscular junction, 79, 84
  histochemistry of, 84
Neuron, 62, 63, 69
Nissl bodies, 63
Node of Ranvier, 65
Nucleolus, 5
Nucleus, 5

O

Oligodendrocyte, 67
Oocyte, 160
Operculum, 96
Osteoblast, 43
Osteoclast, 46
Osteocyte, 45
Ovary, 160–162
Oviduct, 164

P

Pancreas
  endocrine, 119, 177
  exocrine, 7, 8, 119–121
Papilla, renal, 139
Pars distalis, hypophysis, 169, 170
Pars nervosa, hypophysics, 172–175
Pericyte, 67
Peritoneum, 103
Peroxisome, *see* Microbody
Peyer's patch, 98
Pineal body, 175
Pinealocyte, 175
Pituicyte, 172
Pituitary gland, 168–175
Plasma cell, 35, 91
Platelet, 46, 48
Podocyte, 130–132

Prostate, 155–157
Protein secretion-milk, 165
Purkinje cell, 69

R

Replica, 22, 111
Reticularis, suparenal gland, 177, 181
Reticulum
  agrandular endoplasmic, 7, 9, 10, 12, 125
  granular endoplasmic, 5, 7, 9, 120, 125
Retina, 196–201
  desmosome layer, 197
  inner plexiform layer, 200
  Müller cell, 197, **200**
  outer plexiform layer, **200**
  pigment cell, 196
  rod spherule, 200
  rods and cones, 196–200
Ribosome, 7

S

Sarcomere, 81
Sarcosome, 82
Satellite cell, 62
Satellite, centriolar, 16
Schwann cell, 5, 65
Sheath, mitochondrial, 149, 150
Sinusoid, liver, 2, 3, 123
Skin, 57, 58
  appendages of, 58
  eccrine sweat gland, 58
  sebaceous gland, 58
Somatotrope, 169
Space of Disse, 1, 2, 123
Sperm, tail, 150, 153
Spermatid, 145–149
  Golgi stage, 146
Spermatocyte, 145
Spermatogonium, 145
Spermatozoon, 150
Spleen, 92, 93
Stereocilia, 153, 201
Stomach, 54, 104–106
Stratum
  corneum, 57
  granulosum, 57, 160
  spinosum, 55
Synapse, 69–72
  axodendritic, 70–72
  dendrodendritic, 70
Synaptic vesicle, 70, 72
System
  circulatory, 95–100
  digestive, 101–116
  female reproductive, 159–166
  male reproductive, 143–158
  urinary, 129–142

T

Terminal, autonomic nerve, 79
Testis, 145–153
Thymus, 91
  reticular cell, 91

Thyroid, 184–186
Thyrotrope, 170
Tongue, 54, 55
Tonofibril, 13, 55, 57
Tonofilament, 55, 57
Tooth, 102, 103
Trachea, 30
Tubule, 102, 103
  connecting, 139
  distal, 130, 139
  proximal, 130, 135
Tuft, glomerular, 130

U

Urethra, penile, 141
Uterine tube, 164

Uterus, 163

V

Venule, postcapillary, 98
Vesicle
  acrosomal, 146, 147
  seminal, 155
  synaptic, 70, 72
Villi, duodenal, 106–109

Z

Zona pellucida, 160
Zonula adherens, 15, 111
Zonula occludens, 15, 111